The Superintendent's Leadership in School Reform

Dorothy Fast Wissler
San Bernardino Valley College
and
Flora Ida Ortiz
University of California, Riverside

The Falmer Press

(A Member of the Taylor & Francis Group)
New York, Philadelphia and London

UK   The Falmer Press, Falmer House, Barcombe, Lewes, East Sussex, BN8 5DL

USA   The Falmer Press, Taylor & Francis Inc., 242 Cherry Street, Philadelphia, PA 19106–1906

First published 1988

Library of Congress Cataloging-in-Publication Data

Wissler, Dorothy Fast.
    The superintendent's leadership in school reform.

    Bibliography: p.
    Includes index.
    1. School improvement programs.   2. Schools — Decentralization.
3. School superintendents.   I. Ortiz, Flora Ida.
II. Title.
LB2822.8.W57   1988   379.1'535   87–20072
ISBN 1–85000–261–4
ISBN 1–85000–262–2 (pbk.)

Jacket design by Caroline Archer

Typeset in 11/13 Bembo by
Imago Publishing Ltd, Thame, Oxon.

Printed in Great Britain by
Redwood Burn Limited, Trowbridge, Wiltshire
and bound by Pegasus Bookbinding, Melksham, Wiltshire

Contents

Contents

Chapter 1

The Superintendent and School Reform

The mention of the office of the superintendent has been noticeably absent in the current school reform efforts. The reports, starting with the *Nation at Risk* (1983) on through the Carnegie Task Force Report (1986), California's Commission on the Teaching Profession (1985) and the teacher education reforms proposed by the Holmes Group (1986) and the Governor's Commission (1986) have focused almost exclusively on school site reform. (For exceptions see Cruz, 1985; Bacharach and Conley, 1986; Goodlad, 1985). This preoccupation with school site is in direct contrast to the reform movement in the late 1960s and early 1970s. At that time, the efforts were directed at decentralizing school districts with the belief that the superintendent's office required reform. Those districts which were heavily involved in decentralizing resulted in superintendent turnover. (See, for example, Peterson, 1976; Gittell, 1967a and b; Sizemore, 1981). There was an attempt during that period to hold superintendents responsible for their district's reform efforts.

Another contrast to the current educational reform is the organizational reform movement in the business and corporate segment. The literature reports almost exclusively a preoccupation with the executive levels. For example, Ouchi (1981), Peters and Waterman (1982), Deal and Kennedy (1982), and Schein (1985) place the burden of comprehensive organizational reform on the executive posts. However, as stated above, the educational enterprise is presently focusing on teachers and principals.

An examination of successful reforms and the superintendency serves to demonstrate the importance of considering the superintendent's role in educational reform. The present work is a case study of a superintendent who set out to improve the school district he led. The reform took place during the period of decentralization, but this superintendent did not begin with decentralization, rather, in the process of reforming and improving the school district, its structure was changed from a highly bureaucratized organization to a decentralized effective one. This case study demonstrates the critical role the superintendent plays in school reform efforts.

A pervasive point raised in the majority of the reform efforts is the need to grant principals and teachers greater autonomy, professional responsibility, and acknowledgement. For example, Goodlad (1985) and Miller (1986) both advocate reducing rather than increasing district-wide programs and demands, and giving more rather than less autonomy to principals and teachers. Additionally, it is suggested that contextual as well as outcome criteria be used as measures of successful school performance. These authors do not address the issue of decentralization in order to accomplish these aims. Nevertheless, from different perspectives, Chapman and Boyd (1986) and Cruz (1985) support decentralization as a means for bringing about school site compliance within a school district.

This work addresses the issue of decentralization in three major ways. First, it presents a case study of the process of decentralization in a medium-sized urban Southern California school district. Second, it demonstrates how decentralization affected the performance of students and school personnel as well as the quality of work life for all. Finally, the nature of the decentralization of a school district is extensively described. The importance of organizational, structural, and behavioral changes is highlighted as the transformation of an organization's culture is portrayed.

Theoretical Framework

This work draws from leadership theory, organizational theory, change theory, and decentralization research. The analysis establishes that intentional leadership style is required for organizational change which transforms a bureaucratic institution into a decentralized one through seven stages. Successful change relies on the intentional leader's control of the technological core of the organization — information — through three critical points. The organizational cultures present in both the bureaucratic and decentralized organizations are identified through the quality of participation of the organizational members, the use of metaphors and the achievements of both the school personnel and students.

Leadership theory has contributed to a prominent model emphasizing the 'vertical dimension of authority and dependency, of superordinate and subordinate, of decision maker and implementor, of master and servant' (Lippitt, 1984, p. 363). As Lippitt explains, the present challenge to leadership is lodged on the implications arising from the structure described above and the pride in independence and autonomy rather than pride in interdependence and mutual help, and the belief that competition is a necessary and needed support for motivation to achieve. To address the challenges facing present-day leaders, the current focus in the study of organizational leadership has been on the cultural characteristics of organizational life. Schein (1985), in particular, has linked leadership attributes and organizational characteristics to a cultural analysis. This analytical approach has served to de-emphasize the vertical dimension associated with bureaucracies.

Related to schools, Astuto and Clark (1986) explain that effective school leaders 'concentrate on establishing distinction through enactment processes rather than projecting or defining intent in task forces on goals' (p. 62). They 'foster a sense of individual efficacy and *espirit de corps*', place the participants in responsible positions, and set the 'stage for them to invest their energies and skills in the organization' (p. 65). This view accentuates the value of lower participants (Etzioni, 1961) and establishes the leader's responsibilities in relation to them.

Winter (1966) and Schutz (1967) present the interdependent nature between persons, cultural characteristics and social responsibility. The concept of intentionality refers to the responsible nature the leader assumes in directing an organization for the common good.

The process of decentralization has been predominantly investigated through the imposition of decision-making models (Allison, 1971). This conception has emphasized the struggle between self and competing interests rather than the issues of the common good. In this work, the process of decentralization is determined and seven stages are identified. The first stage is the conception of schooling processes by Mr Berry, associate superintendent at the time. The second step includes the discussions Mr Berry held with the president of the school board and the then superintendent concerning schooling and the improvement of education for all. The third stage entails the reduction of the central office and changing it from a command to a service unit. The fourth step is to form a cadre of managers, or supporters for the superintendent. The fifth step grants principals budgetary and programmatic autonomy. The sixth step creates clusters made up of teachers, principals and the superintendent.

The seventh step is recentralization in order to avoid anarchy and the creation of school site fiefdoms.

The process by which decentralization took place accounted for the quality of participation of the organizational members. The school personnel were challenged to improve their work performance and the students were stimulated to achieve academically. The success, however, of changing from a bureaucratic to a decentralized organization is due for the most part to the control of the technological core — information — available to organizational leaders. Kanter (1983) refers to information as one of three basic commodities. Glasman (1986) claims that the moment school leaders begin to process information and the moment they begin to develop a rationale for a potential decision or action is critical to leadership. Information, and most crucial the control of information, is the technological core for organizational leaders. This is simply stated by Glasman when he points out that:

> the territory which has to be charted lies between the moment school leaders begin to process information and the moment they begin to develop a rationale for a potential decision or action. The timing of the first point can probably be adjusted if leaders are able to control when they begin to intentionally process information. The timing of the second point can probably be adjusted if leaders are able to control the effect of external pressure to develop a rationale for their decisions or actions. The more they are able to stretch the territory between these two points the better their judgment may be (p. 160).

The control of information contributes to successful change processes in two different ways: to reduce conflict and a way to determine decisions and actions. In the management of school organizations, three points of information control are necessary: the point where student data are gathered, processed and transmitted; a point where educational theory and data are analyzed, processed and transmitted; and a point where local and social data are collected, compiled, analyzed, processed and applied. This function is so critical and sensitive that direct control by the superintendent is imperative. In the present case the superintendent collected the information, analyzed it, made choices regarding it (intentional behavior), had complete authorization over it, and transmitted or executed actions and decisions consistent with the nature of the information (see Mintzberg, 1979).

The process in the control of information begins at point 1 with the collection of data. The leader starts to filter the information through the core value of community or common good of the organization. Three fundamental assumptions particularly relevant to school leadership underlie the core value: (1) children will learn; (2) those engaged in the delivery of educational services are responsible for what is learned; and (3) the quality of the participation of organizational members matters. The subsequent decision or action taken at point 2 is conditioned by the core value and assumptions presented above. The resultant consistency between decision or action and value system serves to control the effect of external forces.

The consistency provided by a one-person control of information, by the intentional style of leadership and the concern with process provided for a cultural change of an organization from a bureaucracy to a decentralized institution in which challenge and excitement pervaded over the school personnel and student body. The results were high morale and improved student achievement.

The second part of this chapter details the research methods used to collect the data for this study.

Research Methodology

Research Design

The study under investigation took place between 1960 and 1978. It was also during this period that the greatest amount of decentralization demand in the United States took place. Another reason for covering this period is to allow a comparative study of the district under two superintendents. It is a basic study and an historical analysis.

> Organizational theorists in particular, tend to ignore the historical roots of a social system ... The case study technique, coupled with serious attention to organizational history does, however, provide a depth and richness that highlights many problems of organizational change (Baldridge and Deal, 1975, p. 429).

Riverside is a school district which is well-known and has a high profile worldwide because it has been the focus of much research on desegregation. Several scholarly writings are available which differ from the readily-available media material found for most districts. Three of these studies are Hendrick's (1968) history of the integration plan, Mercer's (1968) work on issues in desegregation, and Gerard and Miller's (1975) research on integration outputs.

Riverside experienced the same major educational concerns as other cities around the nation. While other cities responded in different ways, Riverside's response was to radically change its structure from a bureaucratic organization to a decentralized one. This change was due to the superintendent's efforts.

Data Gathered

Interviews

Data gathered for this study consisted of information from interviews, documents, and observations. Data were gathered from two types of interviews: retrospective interviews in which respondents were asked to describe the changes brought about by the superintendent as they recalled them and interviews having to do with the respondents' retrospective analysis of actions taken between the dates under study. The former were interviews with persons who lived through the change process and were in some way a part of it. The latter were, or may have been, persons who did not take part in those events, but who lived with the results. The first provided information on implementation, and the second on the impact.

Many of these interviews were what is considered oral history. 'Even the severest critics of oral history admit that the method works best for gathering impressions, opinions, and attitudes rather than for determining fact . . . (But) if we fail to interview . . . (we) treat that man as a statistic rather than an historical actor' (Harney, 1979, p. 2). No amount of research will allow a full viewing of the process or the impact, but the interview is a vital portion of the saga, providing personal perspectives on the phenomenon.

Names of those to be interviewed came from the district board minutes, the local newspapers, and from participants in the process. Certain names were prominent in the minutes and in the newspaper during the years under study. From these sources a list of 30 names was made. The first person interviewed was E. Raymond Berry who was the superintendent at the time. As the first and chief informant, he was asked to add to, and to delete from, the list in order to include those most instrumental and involved in the process. Other persons also added names to the list. Of the original list, three persons had died and one was very old and not fit enough to tolerate an interview. Several had been with the district only at the beginning of the effort and had left for other jobs and one person could not be located by his father. No one refused to take part in the study when requested to do so. The final sample contains a selection of persons who were involved in or knowledgeable about the school district changes brought about during the period of this study.

Through interviewing it soon became obvious that the process of decentralization in Riverside was a process of structural change in the school district. There were enough interviews with community and school persons which, while representing different views, confirmed that the school district had changed under the direction of Mr Berry, the superintendent. No one demanded decentralization, yet teachers, parents and community groups were prime recipients of the changes which took place. Thus, many participants were affected by the changes from bureaucracy to decentralization.

The final interview list included:

Present position	Past positions in Riverside Unified School District (RUSD)	Years in district
1. Community member (Male)	Mexican-American parent, children in RUSD during the decentralization period. Served on a Community Advisory Group.	
2. Community member (Female)	White parent who had children — most of whom went through RUSD during the decentralization period. Active in PTA.	
3. Community member (Female)	Black parent who had children in RUSD during the decentralization period. Became active in compensatory education committee for the district.	
4. Lawyer (Male)	White lawyer — member and president of the RUSD Board of Education	1962–1976
5. High school principal in another school district (Male)	Black 6th grade teacher in RUSD Middle school principal and high school principal	1960–1961 1968–1976
6. Retired (Female)	White director of research and testing	Before 1960–1977
7. Elementary principal (Male)	White, active as an elementary principal during the period of study and continues as principal	Before 1968–1983
8. Elementary principal (Female)	White special education teacher	–1976
9. Administrative assistant to superintendent (Male)	White teacher, junior high math junior high vice principal junior high principal Director of secondary education	1950 1956 1959 1962
10. Assistant superintendent personnel (Male)	White teacher, junior high Curriculum consultant Assistant director, personnel high school principal administrator of instruction manager of employee relations Assistant superintendent, personnel	1959 1962–1963 1964–1965 1965–1970 1970 1973–1975 1976–1983
11. Assistant superintendent, business (Male)	White comptroller, Assistant business manager Assistant superintendent, business	1962 1983

Present position	Past positions in Riverside Unified School District (RUSD)	Years in district
12. Associate super- intendent, acting superintendent (Male)	White junior high teacher, science junior high counselor junior high vice principal	1957
	Personnel assistant	1965–1966
	Administrative assistant	1968–1969
	Assistant superintendent	1969–1971
	Associate superintendent	1971–1983
13. Lecturer in education university level (Male)	White personnel director	1960
	Assistant superintendent, instruction	1961
	Associate superintendent	1962
	Superintendent-retired 1978	1968–1978
	Lecturer in education	1983
14. Superintendent of schools in California district (Male)	Mexican-American administrator	1977–1979
	Administrator of instructional support services	1980–1981
	Superintendent of schools	1983
15. Teachers		
16. Students		

The sample included three ethnic groups, both sexes, and some differences in education and socioeconomic status. Informants worked in the central office, schools, homes, and in the community. One was totally opposed to the changes which took place. One favored them at the time and became opposed to them during the process. Three opposed some portions and favored others. Several wished to be considered neutral. Twelve spoke in glowing terms and strongly supported decentralization and the changes which took place. It is obvious that the sample contained different philosophies, opinions, and positions.

The shortest interview was one hour. The longest was three and one-half hours. Most first interviews were one and one-half hours. Some persons were interviewed three times, some two times, and some only once. Repeated interviews were for clarification or because interviews were interrupted and not completed. All interviewees gave permission for second interviews. Each person was contacted by telephone before the interview. A careful explanation of the study was given and an appointment was scheduled. Permission was obtained from all but two people to tape record the interview. Copies of the transcriptions were mailed to those who requested them for any additional comments they wished to make. Additional comments were made by two persons. No person deleted pertinent material from the transcriptions.

The interviews resulted in 196 pages of single-spaced typed data. The data were placed on note cards under categories, and the note cards were filed according to subject. Some cross-indexing was necessary. The process of categorization was aided by the subjects of the questions which had been asked during the interviews.

The process of interview generally followed the detailed guidance provided by Metcalf and Downey (1977) for historical studies. They emphasized that the interview must be carefully planned after written materials have been gathered and analyzed. Gathering documents first helps determine the kinds of questions the historical researcher wishes to emphasize. They suggest: (1) organization of questions into an outline form for use in the series of interviews; (2) starting with general noncontroversial questions; and (3) limiting the interview to a maximum of one and one-half hours.

As closely as possible a basic group of the same questions were asked of each person.

1   How did you first hear or read about the word decentralization?
2   In your opinion, did a process called decentralization take place in this district?
3   If it did, why do you think it took place here?
4   How do you define the word as it happened in this district?
5   What factors or forces outside the school district caused it to happen?
6   What factors inside the school district caused it to happen?
7   What steps were taken to put it into place in the district?
8   What persons or groups opposed the process?
9   What persons or groups approved the process?
10 What was your part in the process?
11 How did the district differ after the process from the way it was before?
12 What did the process change?
13 Who made what decisions before and who made what decisions after the change?
14 What were the negatives and what were the positives you found about it?
15 Where is the district now in the process of decentralization?
16 In your opinion, how does Riverside differ from other districts (where you have been employed, where you have friends working, or that you know about in some way)?

Each person was given an opportunity to discuss the change process which took place while Mr Berry was superintendent. Since each person was generally asked the same set of questions this strategy provided some verification of each person's strength of recall. In any analysis, present or historical, one cannot succeed in capturing the full event, for no one person sees it in its totality. The data and data sources were verified and contrasted to reconstruct the history of the event as closely as possible.

Documents

The second major method of gathering data was the collection of documentary material. The major sources of document collecting were the Riverside Unified School District office, the Riverside County Schools office, the press, and personal, private papers.

Two types of documents are not included: (1) routine, day-to-day ditto-type material which is not kept; and (2) personal and sensitive correspondence. Seven types of documents were used for this study: (1) internal organizational communications; (2) organizational communication with external groups; (3) communication with parents and parent organizations of the schools; (4) local and other newspapers; (5) board minutes; (6) budgets; and (7) miscellaneous research reports, books, letters and personal papers. A total of 154 pages of board minute notes were carefully read, analyzed, and categorized. Portions dealing with the decentralization process were tape-recorded, typed, categorized onto file cards and filed according to subject.

The city newspapers were scanned through the use of the *Press Enterprise* clipping file and through the microfilm collection at the University of California, Riverside. Some school district newspapers were also read. Numerous articles were photocopied. Notes were taken from others. A total of 66 pages of notes was made and indexed.

Mr E. Raymond Berry, the superintendent, permitted unrestricted use of his personal files concerning the district. These files contained private and public communications with the board, district officers, teachers, parents, and others concerning the district. Clearance was sought before there was any publication of material from the confidential correspondence.

District personnel also provided documents concerning the process. Computer searches of relevant literature were conducted through the Educational Resources Information Center, the Social Science Citation Index, and the Sociology Abstracts. Many persons were generous in providing titles and actual copies of old and new books on the subject. Some books and papers were lucky finds in the library or miscellaneous files. This variety of documents provided data triangulation.

Observation

The third major data collection technique was observation for one entire quarter (Fall 1982) in a class which was taught by Mr Berry. The class was titled, 'The School Principal'. The class observation served to determine the relationship between the role of superintendent and the process of decentralization. How does a former superintendent teach and coach some actual (and some aspiring) principals and superintendents? It was a method of verification. Did what was being taught in class match what others reported he had actually said and done in real life in the position of superintendent? The class observations provided data on theories and practices which might have been used in the district. The subject had given permission for the observation without being aware of the specific research question. There was, therefore, no opportunity for lectures to be changed.

Each three-hour class was tape-recorded and transcribed into 184 pages of single-spaced notes. Material pertinent to the decentralization process in the district was re-written on cards under subject headings and filed. It, too, needed some cross-filing. Such an observation added a type of unusual strength in the cross-verification of materials from interviews.

Additionally, another university class, 'The superintendency', taught by Mr Berry was video-taped for analysis. Triangulation was again applied to the various types of data.

The major strategy of data analysis was comparative analysis and the analytic induction of Denzin (1977). Denzin saw analytic induction as an attempted synthesis of all models of analysis. A continued sorting of information from the catagorized note cards into groups of evidence was used to triangulate the data to establish the relationship between leadership style, decentralization, and organizational cultural change.

Limitations of the Research

Each preceding section of this methodology detailed some precise limitations of this study concerning interviews and document collection. In the area of interviews, general limitations included the problem of remembering events which happened a decade ago. Most of the interviewees expressed concern that they wanted to be certain to remember as accurately as possible but 'that it has been a long time' or 'I haven't thought about that for a long time'. One person showed some definite dimming of memory as the interview proceeded. While this is rather easy to spot in some cases, in others it is far more subtle and, therefore, one must be constantly alert. Tape recording the interviews allows full attention to be given to the informant and to the informant's ease or possible confusion. It also allows for accurate comparison of document material later.

The bias of the researchers was in the direction of thinking that the district had undergone decentralization prior to the study. This bias was continuously examined and acknowledged as the interviews, collection, and analysis proceeded. It was particularly crucial that questions not be 'loaded' and that informants not be 'fed' an attitude or a direction. A bias can be both a strength and a weakness. Its acknowledgement was necessary if it were to be a strength. In this case it alerted the researchers to look carefully for opposite case remarks or even hints of remarks.

Time is a limitation of all life and all activities in it. It is particularly so when one is asking for the time of others in interview and evaluation. Several of the individuals interviewed were between heavy travel schedules, heavy desk work and full calendars.

The district had moved its headquarters from a large to a much smaller office during the process of decentralization eliminating the district librarian. That complicated finding some documents that were not essential but would have provided saturation of document data. Nevertheless, the documents obtained were adequate and provided considerable detail regarding the school district's changes during this period.

> There are further general limitations of this type of study. Intensive study of a single organization has its theoretical and methodological dangers. It is easy to be so captivated by the unique events and actors that the resulting explanation has no value for any other organization. But we believe the advantages of a careful case study outweigh the disadvantages. A case study provides developmental perspective on an organization that a cross-sectional analysis of many organizations can never offer . . . A case study also encourages the analysis of contextual effects on an organization and permits the identification

of many kinds of variables and the analysis of interaction among them (Sproull *et al.*, 1978, p. 8).

This work, thus, presents an historical analysis of a superintendent's efforts to improve and reform the organization for which he was responsible. In this process, the organization was changed from a bureaucratic to a decentralized one. The presentation is organized into seven chapters.

First and foremost, superintendents are the primary and most important persons engaged in comprehensive school district reform efforts. The second chapter in this book details how this primacy is established. Second, reform efforts take place as processes which consist of components and characteristics which either enhance or inhibit success. The third chapter presents the process by which the school district was changed from a bureaucracy to clusters of cooperation, the process of decentralization.

Third, specific changes took place as a result of decentralization. The fourth chapter delineates these changes. Fourth, organizational and leadership studies have been consistent in demonstrating the interdependence between leaders' attributes and organizational factors. The fifth chapter deals with the critical organizational factors in this particular school district which interacted with the leader's intentions. This chapter demonstrates how the control of information and the leader's intention interact in an effort toward congruency in organizational behavior. A leader's acuity in understanding these dynamics increases the likelihood of success.

Fifth, several important theoretical constructs have dominated the analysis of school organizations. The sixth chapter demonstrates how theoretical assumptions guide research and practice and how the present case challenges present notions regarding leadership and organizational structure. This chapter illustrates how leadership style, organizational change, and the nature of decentralization interact in new and different ways.

The seventh chapter addresses policy issues related to school reform and the role of the superintendent. Organizational reform and improvement are directly dependent on the executive leader's intentions and actions.

Chapter 2

The Establishment of
Intentional Leadership

Organizations' cultures and leadership styles have been reported in many different ways. This work analyzes how a superintendent transformed a school district from a bureaucratic centralized culture into one of excellence in the instruction of all children, thus showing how leadership style is interdependent with organizational culture. In this particular case, the leadership style is called intentional, but may be likened to the leadership style described by Schein (1985) as the *culture manager*.

In order to understand how school reform starts with the superintendent, it is important to present the concepts which are being applied to explain his leadership role. Intentionality is the core concept used to describe how Mr Berry, the superintendent, led a medium-sized urban school district through school reform.

Menges (1977) begins his discussion of intentionality by asking, 'Why do people do what they do?' (p. 96). In his study of intentional behavior (primarily focused on teachers) he concludes that people *do* what they *know*, what they *like*, what they *can*, what they *must*, and what they *intend*. The study of the superintendent, Mr Berry, shows that intentional leadership is bound by certain characteristics which can be best presented by reviewing the psychological strands which have contributed to the understanding of intentional behavior.

Three schools of psychological thought have contributed most heavily to the definition of intentionality. The psychoanalytic school argues that intention incorporates both conscious will and unconscious wish. May (1969), for example, states that intention and action are inseparable, representing a unity of thought, will and action.

Social psychologists explain intention as the probability of a given action. This model separates the contribution of attitudes and subjective norms and suggests that intentions are altered by changes in attitude in a particular situation and by changes in subjective norms as new information is received.

A third school of thought, behavioral psychology, defines the intentional individual as one who can generate alternative behaviors in a given situation and can approach a perceived problem from different vantage points as environmental feedback is received. Ivey (1969) suggests that the intentional individual is not bound by one course of action, but can act in a moment to respond to changing environmental situations. Alschuler and Ivey (1972, p. 54) define intentionality as 'the process of fusing conscious consideration of alternatives with positive action'. Mr Berry, in his role as school executive chose from several alternatives and acted consistently and congruently with them as he improved the organization.

The philosophic antecedents to the concept of intentionality serve to better explain its potency in the present analysis. Mead (1934) and Schutz (1967) are concerned with the manner in which individuals are able to generate meaning for themselves and explain social meaning. Mead presents the mind and self as *social* emergents with language as the mechanism by which this takes place. Communication, in this framework, assumes primary importance because it is through communication that meaning emerges. Beyond that, taking the role of others, or the emergence of 'me' enables reflective meaning. The complete self contains the 'I' which is capable of creativity in addition to reflection.

While Mead (1934) views the structuring of the social world through socially-generated gesture and response, Weber (1957) sees the social world as interpreted through the subjectively intended meaning of human acts. For Weber, it is only through understanding of individual action that social science can gain access to the meaning of each social relationship and structure.

Schutz (1967), on the other hand, sees the problem of meaning as lodged in history. The meaning of a person's experience is constituted for him as he 'lives through' that experience. We are conscious of action only if we contemplate it as already over. We project the intended action as if it were completed.

The *meaning* of an action is this completed or projected act. Foster (1983) summarizes: 'Action is projected behavior which is imbedded in the intention (intended meaning) of the actor, and this intention is generated by our ability mentally to picture what we are going to do, as if it had already been done. This cognitive map which is consulted distinguishes intentional acts from unconscious behavior' (p. 53).

Schutz (1967) further suggests that individuals come to understand the meaning of both self and contemporaries as they reflect on simultaneously 'lived through' experiences and then construct 'ideal types' out of this understanding. The 'ideal type', while firmly grounded in individual experience, nevertheless enables projection beyond the realm of the contemporary, so that an understanding of the 'generalized other' may be gained. Only through this 'ideal typi-cal' projection is the individual able to acquire the meaning of those outside his contemporary experience. It is this latter characteristic which distinguishes intentional leaders from others.

Menges (1977) defines intentionality as the process of consciously considering alternatives and selecting the appropriate course of action from an array of skills. For the purposes of the present analysis, intentionality is applied to mean that 'individuals can choose to be momentarily aware of the entire physical and social space in which they find themselves; they can also determine the duration of aware-ness as well as choose as the object of consciousness a single object such as the dress of a solitary individual' (Bowers, 1984, p. 43). In sum, the concept of '*intentionality* points to the fact that consciousness is involved in acts of interpretation, imagination and remembrance' (p. 78).

Additionally, the motive underlying intentionalism can be likened to what Winter (1966) refers to as 'cultural integrity'. He writes:

> The issue is one of cultural integrity — of actualizing rights which are implicit in the social order. He may sense that the interests of a submerged group need to be recognized and extended. Or, he may believe that new values need to be apprehended and embodied in new structure for the society (Winter, 1966, p. 284).

Mr Berry's superintendency was intentional because he chose to im-prove the school district. He put it this way:

> It was just my concept that there were so many changes coming, and that our society was emerging so rapidly that there was no way that a large organization could be directed in that manner (top down) effectively. For me it was almost a guarantee of mediocrity because you had to wait for the abso-lute direction of individuals in charge in order for something to happen. And I didn't see how a diverse school system with that many schools with that many differences could be led effectively from a central position.

What were the changes that needed to be brought about in the Riverside Unified School District in order to produce cultural integrity? Three major factors were instrumental in the demand for changes in the school district: urbanization, classroom integration, and quality education for all. We will look at each of these in turn in a general way and then examine each of the three in more specific terms in Riverside.

Urbanization

Large school systems are decentralized more often than the school systems in smaller cities. In 1960, seven out of ten of the largest cities in the United States were decentralized to some extent (LaNoue and Smith, 1973).

Decentralization was 'born' in New York, our most urbanized city. Decentralization law in the State of New York emerged after a legislative struggle in which an uncommitted center finally effected a compromise. The new legislation which was hammered out over a three-year period contained 64 pages of detail in which 31 new community school boards were delegated specific controls over expenditures, personnel, and curriculum (LaNoue and Smith, 1973).

Michigan's decentralization law followed shortly after New York State's in August 1969. The Michigan legislation was only one page long, but it also created regional school boards with some power to control budget, curriculum, and personnel. The common factors Detroit and New York shared were that they were large, urban, had highly centralized school systems which were suffering from a shrinking property tax base, White flight, and high Black unemployment. The cities were unlike in some ways, too. Detroit had been sensitive enough to its Black population to develop textbooks depicting Blacks. It had also suffered a week of riots in the 1960s which killed over forty persons. New York had not rewritten its textbooks and had experienced boycotts, demonstrations, and sit-ins, but not the 'war' of Detroit (LaNoue and Smith, 1973, p. 115).

Since size and urbanization seem to be correlates of decentralization, it is necessary to trace Riverside's history of population growth and urbanization. Riverside did move from an agricultural economy toward a service and industrial economy and became urbanized to the point of developing a blighted central core (Greater Riverside Chambers of Commerce, 1983, p. 96).

Riverside is located in the southwestern part of the United States. It was founded in 1870 and incorporated in 1883 (Greater Riverside Chambers of Commerce, 1983, p. 39). Three men were prominent in its earliest history. Don Juan Bandini was given a large Mexican land grant in 1838 which adjoined the Santa Ana River. Bandini, in turn, gave the rancho called El Rancho Jurupa to his son-in-law, Abel Stearns. Louis Rubidoux bought a nearby section and lived there as a rancher until his death in 1868. After his death, Louis Prevost bought the land to use in the raising of silkworms but he lived only two years, and the land was next purchased by 'Judge John W. North and his delegation of Easterners...on September 13, 1870' (Hendrick, 1968, p. 24).

North held his interest until 1874 but a circular he and the delegation published in March of 1870 called 'A Colony for California', provides an insight about the fledgling colony that perhaps nothing else can give:

> Appreciating the advantage of associated settlement we aim to secure at least 100 good families who can invest $1000 each in the purchase of land; while at the same time we earnestly invite all good, industrious people to join us who can, by investing a smaller amount, contribute in any degree to the general prosperity... Experience in the West has demonstrated that $100 invested in a colony is worth $1000 invested in an isolated locality. We wish to form a colony of intelligent, industrious, and enterprising people, so that each one's industry will help to promote his neighbor's interests as well as his own... We expect to have schools, churches, lyceum, public library, reading-room, etc., at a very early day, and we invite such people to join our colony as will esteem it a privilege to build them (Patterson, 1971, p. 19).

The circular was noteworthy because it spoke of cooperation as a way of life and it listed schools as a first interest of Judge North and his delegation.

According to Hendrick (1968), minority people lived in Riverside from the time of its founding. First, Mexican immigrants and later Mexican-Americans were laborers in the citrus industry. When the Southern Pacific Railroad was laid through Colton in 1876 and the California Southern Railroad was built through Riverside in the 1880s, Chinese provided the labor. When the first Blacks came to Riverside by 1879, most worked for the railroad in San Bernardino (Hendrick, 1968).

One of the original colonists, Eliza Tibbets, was given three Brazilian mutant seedless orange trees which she planted in her yard about 1875. Pieces she gave away were grafted and 'by 1877 the navel orange had become the single most important crop of the growing colony' (Greater Riverside Chambers of Commerce, 1983, p. 5).

In 1893, the growers in and about Riverside, called the Pachappa Orange Grower's Association, cooperatively started what was to become the Southern California Fruit Grower's Exchange. The navel orange farming brought new demands, and as a result the Food Machinery Corporation, Riverside's pioneer manufacturing industry, began production of citrus machinery in the 1880s (Patterson, 1971).

Before World War I, Riverside was largely an orange grove which contained some population (Patterson, 1971). 'In 1907 the Regents of the University of California authorized the leasing of lands in Riverside for a citrus experiment station' (Hendrick, 1968, p. 27). March Field was opened in 1918 and trained the Rainbow Division fighting groups. The community went through World War I with 'no dramatic' growth (Hendrick, 1968).

During World War II, 'Riverside underwent such a gigantic growth of population that houses could not be built rapidly enough' (Greater Riverside Chambers of Commerce, 1983, p. 10). March Field expanded; Camp Haan was built, and eventually housed 80,000 persons; and a third base, Camp Anza, was constructed. 'Between 1910 and 1950 the population grew from 15,212 to 46,764. Ten years later it stood at 84,332' (Hendrick, 1968, p. 30). In 1970, the total was 140,089 and by 1979 it had increased to 162,800 (Greater Riverside Chambers of Commerce, 1983, p. 38).

While all groups increased in numbers during this time, census figures on changing ethnic composition are not particularly helpful since the same designations are not used for each census. There was an increase from 4.9 per cent to 5.3 per cent non-White in the city from the 1950 to the 1960 census, but there is no category of Mexican-American (US Bureau of the Census, US Department of Commerce Publication, County and City Data Book 1967, p. 474). 'By 1960, Blacks constituted 4.7 per cent of the total population and Mexican American 8.5 per cent' (Gerard and Miller, 1975, p. 27).

The growth of the minority population from 1965 to 1974 in the city schools increased so that between 1965 and 1973 Black students increased from 6.09 per cent to 10 per cent and Mexican-American students increased from 9.96 per cent to 15 per cent of the total enrollment (Gerard and Miller, 1975, p. 27).

From the 1880s until after World War II Riverside was an

agriculturally-centered economy. Tom Patterson's (1971) history of Riverside discusses its growth from agriculture. Beginning in the 1950s two brothers, Joe and Ed Hunter, moved from the manufacture of venetian blinds into aluminum casting. By 1970 Riverside had three well-established aluminum industries as a spinoff from these small beginnings. The aluminum industries along with the quantities of aluminum siding they manufactured attracted the mobile home industry to the city.

Bourns, Incorporated, the manufacturer of instruments to guide rockets, located in Riverside in 1950. Twenty years later the company employed 14,000 persons. Rohr Corporation moved into the largest of the Camp Anza deserted wartime buildings in 1952. About the same time W. Atlee Burpee Seed Company joined Rohr in another former Anza building. The Smith-Scott Pipe Company was established to provide steel linings in concrete pipe to increase the life of water systems. In 1958 the Lily-Tulip Cup Corporation built a new plant in the city and in the same decade the *Wall Street Journal* chose the city as a location for the printing of one of its western editions.

By 1960, as this study begins, industries had begun to replace the groves of citrus on Iowa Avenue with manufacturing plants, and the Pioneer Food Machinery Corporation was faced with the necessity to redesign its citrus machinery for use in other industries. In July 1963, there were 8750 employees in manufacturing and only 1600 employees in agriculture, forestry, and fisheries combined (Economic Industrial Site Information, 1965, p. 15). The city actively pursued industry, and on January 5 1965, the City Council unanimously adopted a policy on industrial development which had echoes of the 'Judge John North circular':

> All stable, financially sound, and progressive industries which meet the use standards of the various industrial zones of the City are cordially and aggressively invited to locate their facilities in our beautiful, economically sound, and progressive community (Economic Industrial Site Information, 1965, p. 3).

Such advertisements plus pledges of 'full cooperation' resulted in continued industrial development. By 1970, 'The County Department of Development listed some 75 manufacturers in Riverside, other than those who...produced primarily for the local market' (Patterson, 1971, p. 436). While the total number of manufacturing establishments in 1958 was 78, in 1963 it was 101, and in 1967 the number had risen to 131 (US Bureau of the Census, US Department of Commerce

Table 1   *Riverside, California Registered Voters, 1960–1978.*

Year	Registered Republicans	Vote	Registered Democrats	Vote
1960	21,109	19,311	19,284	17,063
1962	22,875	18,387	21,881	16,647
1964	27,514	23,146	29,664	25,074
1966	28,611	22,464	30,018	21,309
1968	28,919	23,903	30,421	24,485
1970	26,252	21,052	29,160	23,166
1972	30,279	23,871	40,709	31,282
1974	27,677	18,533	39,287	24,720
1976	26,418	22,623	40,096	32,232
1978	26,747	20,681	40,498	28,371

Source: Riverside County Registrar of Votes — General Elections.

Publication, County and City Data Books, 1967, p. 477; 1972, p. 650). The city seemed a unified one to be able to marshal its resources to pick up the slack from agriculture into industry. For just such vision the city was given one of the eleven 'All-American' cities awards in 1955. It cited Riverside's ability to take action before its problems became acute in the areas of 'explosive growth', future water needs, and city government (Hendrick, 1968).

By 1970 citrus growers were troubled. Fruit was growing smaller and the total crop output was reduced. Smog and water pollution were blamed. Land values were increasing and citrus growers were curtailing their production and selling their lands to speculators and subdividers. Now its second-most important industry (manufacturing was first) was receding. In 1970, a modern shopping mall away from the central city opened in anticipation of even further population growth (Patterson, 1971). Industry seemed to be Riverside's future.

Riverside was known as a 'Republican town since its early founding' (Hendrick, 1968, p. 22). As the population grew and the main industry changed from agriculture to manufacturing, the number of registered voters changed from a Republican to a Democratic majority (see Table 1).

The stability of the city is reflected in the fact that it had only 16 mayors from 1907 until 1978 and forty city councilmen from 1951. Seven of the forty were in office in 1983 (Riverside Mayor's Office, 1983).

Since 1952, Riverside has had a council-management form of government with a seven-member council elected from geographical wards for four years. The mayor is elected at large every four years and presides over the council which appoints the city officers —

manager, clerk, attorney — and all boards and commissions. The charter delegates the responsibility for administration of city affairs and the employment of all department heads to the city manager (Greater Riverside Chambers of Commerce, 1983, p. 36).

By 1970 urbanization with its suburban sprawl had proceeded to a point where business was following the spread of population and was leaving the downtown area. The opening of the Tyler Mall with the May Company, Broadway, and Penny's Department stores plus smaller retail shops in 1970 far from the center of Riverside signaled the city council that there was a need to revitalize the downtown area. It adopted a Redevelopment Plan in 1971 which was expanded to include 125 blocks in 1975 in order to remove, rehabilitate, and modernize buildings and to upgrade public services and facilities in the area (Greater Riverside Chambers of Commerce, 1983, p. 94).

Though the population was spreading out with urbanization and sprawl, Riverside's leadership wished to 'retain the community's rich cultural and historic background while at the same time bringing the downtown area to its full potential as a government, professional, and financial center' (Greater Riverside Chambers of Commerce, 1983, p. 87).

The blighted city core was a reminder that the 'character of United States life and Riverside's life changed in many ways during the 25 years following World War II — in its urbanization, in the faster pull toward racial and ethnic separation and the polarization of power and wealth' (Patterson, 1971, p. 466). The changing demographics can be seen as the population grew from 46,500 in 1950 to 146,500 in 1973.

By 1970 the total number of families in Riverside living below the 'low income level' was 2900 (14,645 persons). This was 8.3 per cent of all families in the city, with 36.7 per cent of all the children under 18 years of age in the 'below low income level families' (US Bureau of the Census, US Department of Commerce Publication, County and City Data Book, 1972, p. 646).

Almost one out of ten families and more than one-third of the children under 18 were living in or near poverty in Riverside in the early 1970s. Though Riverside's problems were less than New York's or Detroit's, Riverside suffered from the same urbanization problems in a smaller way.

Classroom Integration

The second large factor contributing to decentralization was failure of school systems to integrate their classrooms. Minority parents had held the hope that integrated classrooms would insure a higher quality of education for their children. In Detroit the school board was committed to integration, but they were recalled and their plan of integration was repealed. In 1966, in Harlem, a joint meeting of EQUAL and the Harlem Parent's Committee reviewed the failure to achieve integration in classrooms in a school called IS 201. This very meeting may have been the last straw which set afire the decentralization calls. The thinking of Black parents shifted that night when a man named Isaiah Robinson said (almost as a joke) that since there was little integration in Harlem, 'maybe the Blacks had better accept segregation and run their own schools' (LaNoue and Smith, 1973, p. 166). From the meetings subsequent to that one came the first proposed comprehensive community control document in New York City. Without integration the best hope for better education for minority children was that their parents should control their schools. Increased demands for decentralization/community control grew directly from IS 201 and its supporters. The mass media picked up the demands and put them on the public agenda.

Detroit had seriously tried to integrate its schools since 1960, but ten years later there were 22 more all-Black schools. In Detroit a school board which was committed to integration was recalled and a legislative mandate repealed the board's integration program. It was replaced by a voluntary attendance plan which was found unsatisfactory. Consequently, in Detroit between 1967 and 1971 significant support among Blacks for integrated schools had declined from 66 per cent to 29 per cent. Many Blacks moved from pressing for an integrated society to preferring community control. In both cities there was a shift in minority thinking from integration to community control (LaNoue and Smith, 1973).

Since 1907, systematic racial segregation of schools had been practiced in Riverside (Patterson, 1971, p. 480). This fact was repeated in most cities across the nation and its history was 'consistent with the national pattern [of] racial and ethnic discrimination in the areas of employment, housing, schooling, and social relationships...' (Gerard and Miller, 1975, p. 26).

In schools, 'separation of Mexican-American children was assumed to be legal and was practiced in the state prior to 1946. Legal segregation of Black children...had not been sanctioned under California law since 1890' (Gerard and Miller, 1975, p. 28). However, a longstanding and deliberate practice of separation had also been inflicted upon Blacks in Riverside. Just as in many other American cities, Riverside practiced a deliberate gerrymandering of its school boundaries in order to exclude minority students from some schools. One school in Riverside which was almost exclusively serving White families changed into an integrated one, and then into a school serving minority families almost totally all inside of ten years (Hendrick, 1968, p. 68). 'The school board perpetuated the administration of segregationist practices from about 1910 until 1952' (Hendrick, 1968, p. 32).

Nationwide there began to be challenges to such long standing practices of separation and exclusion. There were changed interpretations of what constituted justice, landmark legal decisions, and a changed societal climate.

There was a loss of confidence in schools and at the same time there was a demand that meaningful integration take place. For a decade before the Supreme Court ruled in 1954 that segregated schools were inherently unequal, critics had pointed out that segregated schools in a democratic society had not educated all children. There was serious discussion by scholars of the *Equality of Opportunity Report* by Coleman (1966). Its findings were that there was a lateral transmission of values in schools and therefore a measurable achievement gain for minorities in majority classrooms (Gerard and Miller, 1975, p. 3). The addition of this scientific report to humanitarian and social justifications for school integration increased a growing social context which questioned the justice of separate educational facilities.

Legal mandates to end *de facto* segregation became clearer. 'In 1962 the California Board of Education, the state courts, and the legislature began to strengthen markedly the legal case for school desegregation' (Hendrick, 1968, p. 15). The California Supreme Court in 1963 made it the obligation of school districts to detemine school boundaries while being color conscious in order to effect desegregation. Title V, Chapter 7 of the State of California Administrative Code (State of California, 1977, p. 23) established the State Board of Education policy 'to provide equal educational opportunities to all students regardless of race, religion, ethnicity or sex' declaring that 'school districts have a constitutional obligation to take reasonably feasible steps to alleviate the racial and ethnic segregation of minority students'. The State Board also found that 'California school districts should proceed to implement this legal and educational obligation without the necessity of protracted and expensive court proceedings'.

At the same time, the national climate pressing for justice, integration, and rights for all grew in intensity. The election and 'New Frontier' of John F. Kennedy, the passage of the Civil Rights Act in 1964, and the nonviolent demonstrations under the leadership of Dr Martin Luther King, Jr, added to the climate.

The summer of 1964 saw demonstrations and looting in Harlem. In August 1965 the problem moved much closer to Riverside as the Watts riots in south-central Los Angeles left 35 dead (Patterson, 1971). During that time a story appeared in the *Riverside Press* about several Blacks in a Riverside bar who suggested that they 'burn here, too':

> Barely two weeks later...someone set fire to one of the buildings of the Lowell School in the eastside ghetto area. People in Riverside, of all ethnic groups, were generally edgy.... Agitation by minority parents for improved education for their children seemed to be reinforced by the general unrest (Gerard and Miller, 1975, p. 1).

The national unrest had struck in the heart of Riverside. Earlier local headlines had been about 'Racial Violence Erupts in North, South Cities' or 'Mississippi Governor Calls Out Troops to Prevent Riots' (Hendrick, 1968, p. 20). Suddenly, there were headlines in Riverside's own paper about Riverside — it was in crisis.

Besides the national picture and climate, Riverside 'experienced arson which destroyed one of the segregated schools, a petition to the school board calling for integration, and a school boycott. Yet it all happened inside of two weeks and with a minimum of strong feeling generated before or since' (Hendrick, 1968, p. 214). On October 25 1965, Riverside adopted a comprehensive plan for the integration of its elementary schools. What is so very significant is that at the time 'few school systems in the United State — and no system as large — had succeeded in doing more' (Hendrick, 1968, p. v). It was accomplished in the space of seven weeks.

Hendrick's meticulous assessment of the history of desegregation in Riverside makes it clear that before the two weeks of crisis there had been no large-scale effort to push for integration in Riverside. He said that there had been two issues leading to the confrontation which ended in desegregation. They were: (1) disappointment over the results of compensatory education in Riverside, and (2) the way in which minority parents perceived the open enrollment policy to be administered in the district. Past organizational affiliation was not a key to desegregation in Riverside. The town's civil rights groups had not organized to integrate; there had been no formation of *ad hoc* groups for the purpose of integration; the minority community had not passed petitions nor had minority leaders exhorted action. Instead, 'in the course of one week, a few parents, supported by many more, had successfully won the attention of the board' (Hendrick, 1968, p. 121). These were Black parents who acted as single individuals or as neighbors in small groups. There had been no leadership from powerful national organizations or powerful personalities from either within or outside the community.

Riverside experienced indirect local and national forces demanding integration of the same magnitude as other cities. Locally the direct pressure it faced was probably more than some but less than other cities. Any school board or community which wished to drag its feet could do so, and did. By 1968, only two cities in California 'and in the nation with populations exceeding 100,000 [had] adopt[ed] complete racial balance plans' (Hendrick, 1968, p. 211). Yet Riverside chose to integrate its schools in 1965. In 1968, the city extended its racial integration plan to include its junior high schools, and it was 'initiated by the administration without prior pressure' (Hendrick, 1968, p. 220).

Looking back after nine years on the 1965 decision to integrate schools in Riverside, 'one can see that the political process involved in bringing about desegregation in Riverside was direct and relatively uncomplicated, while the historical conditions in which the decision was framed were optimum for decisive action' (Gerard and Miller, 1975, p. 50).

Why were the conditions 'optimum' in Riverside and in only one other city over 100,000 for integration by 1968? Why when there was so much national reluctance to desegregate schools was Riverside so progressive? By saying that 'the Riverside School District...became one of those rare places where the price and other conditions were right for change', Irving Hendrick (1968, p. 220), while undoubtedly correct, fails to take us inside the process to the thinking of the professionals or to the change in the social relationships which promoted a 'right price'. Were the intentions in Riverside a match for those in numerous cities and only the conditions different? A 'price' statement reflects an exchange view of society which is a legitimate view. However, it is scholarly to examine the process from other frameworks, such as an intentionalist framework. The exchange view of society is one of the price must be right. The functionalist view is that the organization must be maintained by some kinds of patterned maintenance even if the patterns must change. The behaviorist view is that biology drives the survival and it is biology which is in command. The intentionalist view is that there is a reservoir of purpose which needs examining in every situation. In this view the actors are not just the creatures of the price, the rational goal of preservation of society, or their own biological drives. The person is not a victim of the situation, rather he can intend to act to change the situation.

> Leadership brings about real change that leaders *intend*. . . . It would be idle. . .to measure the extent and character of social change unless we also examine the intentions of those who make the decisions that were intended to bring about change. Such an examination is necessary if we are to find purpose and meaning, rather than sheer chance or chaos, in the unfolding of events. The test is purpose and intent, drawn from values and goals, of leaders, high and low, resulting in policy decisions and real, intended change (Burns, 1978, p. 415).

By examining leaders' intentions we may capture process and intent.

Desegregation has been well researched in Riverside. It is sketched here because the failure to put it into place in most cities in the nation is considered one of the major influences triggering the call for decentralization and school reform. In New York and Detroit where parents demanded integration, it was not put into place. The demand switched to one of decentralization, and that was largely not put into place either (Fantini and Gittell, 1973). In Riverside where integration in the classroom was quickly instituted and where there was no community demand for decentralization it, too, was instituted. Clearly there was a variable in Riverside which was missing in other cities across the nation which, when faced with either demand, rather successfully resisted both. Therefore, the examination of intent in Riverside seems to be useful.

Quality Education for All

Education crept into the 1960s under the shadow of Sputnik. The Russian feat brought an added sense of failure to Americans who immediately blamed their schools. Therefore, the major educational concerns of this time were how to teach every child and how to change the schools to increase their effectiveness. The pressures for changes came from parents, some teachers, and the federal government. It was not minority parents alone who were aroused by student failure. After Sputnik, middle and upper-class parents, too, were generally demanding an increase in the effectiveness and efficiency of formal instruction to produce higher quality education (Parks, 1959). There was serious discussion by scholars and politicians about the failure of the schools to teach the children from poor families of all races. It was felt that everyone could be well educated if only the system were changed (Schwebeck, 1969).

The humiliation of Sputnik, demands for changes, and federal monies produced an avalanche of innovation. As of 1968 one researcher found more than 1250 new educational practices being tried in the 16 large school systems he investigated (Boutwell, 1969). There had always been pressures on schools and schools had always responded, but during the 1960s and 1970s the pressure was more intense and the responses more dramatic. There were curricular content, organizational structure, and technological changes (Kowitz, 1963).

The 1950s produced a storm of curriculum innovations (Bartlett, 1957). Everything was in question. What was a basic curriculum? Disagreement raged over what was considered a 'core'. By 1961 university professors were writing secondary subject curriculum. Government funding encouraged their efforts. Their entrance produced changes in content with the introduction of 'new math' and 'new science' (Wedel, 1966).

Foreign language was introduced into elementary curriculum as some citizens realized the cultural provincialism almost insured by entrapment in one language only. A second language in the classroom did not raise the storm that the later introduction of bilingual education did. Acrimonious debate surrounding bilingual education raised basic educational questions.

Cross-cultural components were added to curriculum amid conflict. Materials were written in a new 'bowl-of-minestrone-soup' rather than 'melting pot' style (Boutwell, 1969). Mexican-American and Afro-American history were offered in some schools (McEachern, 1968). Curricular changes touched the core of what some parents felt to be old and tried 'basics' and new texts such as *Land of the Free* raised storms of protest. Attempts to look at the role of the self-concept in learning threatened some parents (Landsman, 1962). So did the use of psychometric techniques which some called an intrusion into the privacy of the home.

One of the most basic educational concerns was about reading. There were questions about how to teach it and when to begin to teach it. Was kindergarten too early or too late? Was learning more effective with or without machines? In 1964, *ita*, the 44-letter alphabet was introduced as an improvement in the teaching of reading. By 1970 it had spread across the country as had the controversy about its effectiveness. There was research on whether phonics or the see-and-say method produced more mastery and which of those should start when.

Curricular changes placed stringent demands upon schools and upon teachers' time and energy. However, innovations in organizational structure were just as demanding. The new math and new science encouraged new ways of relating to the student. Brainstorming, creative problem-solving, and the independent study method did not fit tightly into traditional teaching techniques (Hermanowitz, 1959). Boutwell (1968) said that between 1963 and 1968 the federal government spent billions to encourage the development of just such new methods in education. The provision of funds paid off in organizational change which touched all roles in the schools.

In some places change broke lock step learning. A popular plan removed grade barriers (Kowitz, 1963). There were new kinds of grading, non-graded schools, schools without walls, and modular learning, all of which encouraged students to move at their own pace. An open classroom movement was imported from Great Britain. Some schools experimented by having no subjects (Harris, 1972), or having so many subjects that learning in them was dubbed the 'cafeteria concept' (Heidelberg, 1971).

A second organizational change shook the reliance on one teacher lecturing to one class as the only instructional pattern in education. Team teaching, large and small group instruction, and lay instructional aides were introduced (Kowitz, 1963). A second adult in the classroom was a structural change of magnitude. These efforts were aimed at producing more mastery learning by developing more one-on-one time with an instructor. New style buildings were designed and constructed to accommodate these organizational changes. Walls were moveable and carpet appeared everywhere to mute noise.

The pressures of the 1960s also brought rapid developments in technology which were designed in part to free the teacher and the student from the textbook. The effectiveness of instruction was to be enhanced by programmed learning, multimedia, and educational television (Ramsey, 1963; Bright, 1967). In 1962 President Kennedy signed into law a bill providing $32 million for the development of educational television. It resulted in 52 television stations available to schools. Technology both challenged and promised help to teacher and student.

The major educational concern of the times was a nationwide desire to educate children more effectively. From that sprang demands for changes which resulted in innovations in curriculum, structure, and technology in the schools. Pressures on the schools had meant pressures on teachers who had taken the brunt of the attack concerning children's failure to learn. They had responded by innovating and/or accepting the structural and technological changes. Each of these had consumed their energies. They felt pushed around by a bureaucracy in which they felt they had no control or influence. Teacher militancy rose as they demanded a voice in running the schools for which they were in large part held accountable. Their complaint was that they worked for near poverty wages while having no say about the professional rights or their working conditions (Boutwell, 1968). The militant talk became action with a first strike in New York City in 1963.

As the 1960s progressed, parents in large cities found the results they had hoped for were not forthcoming. Drop-out rates were increasing. Detroit's drop-out rate reached 19 per cent, 'which was relatively low', but there was also skepticism about the competency of those who did graduate (LaNoue and Smith, 1973, p. 118). The 2400 federal projects which were set up to educate functional illiterates across the nation were evidence of school failure. Minority parents who had at first only demanded their children be taught to read in integrated classrooms began to demand control of their schools when integration did not fully come to pass.

Poor, middle-, and upper-class parents joined in the demand for quality education for their children when innovation did not relieve the problem of student academic failure. Mayor Lindsay of New York said that this demand was not sudden but that it came from a growing belief that public schools were failing to teach children (Boutwell, 1968). Parents who had begun in the late 1950s by saying that all children needed an adequate education (Parks, 1959), by 1970 were saying that radical reform was needed to make the schools accountable (Fantini, 1972).

When the United States Commissioner of Education, James E. Allen, said in 1970 that every youngster had a right to read and that no child by 1980 in any school anywhere in the United States would leave school without the ability to read properly, he spoke of commitment. The statement also pointed to what some assumed was the continued failure of the 1950s. Not all children had yet been taught to read or were educated properly. This failure spurred the desire of parents and teachers for decentralization in public schools.

Riverside experienced the same major educational concerns as other cities around the nation. While other cities responded largely by implementing pedagogical innovations, Riverside's response to this innovation movement was to decentralize (see Fletcher, 1986).

Intentional Leadership

Mr Berry's superintendency was intentional because he *chose* to improve the school district. One school official described this action: 'Berry was out to break a stifling framework and he ripped it wide apart and he got people creating'. Berry was not a centralizer. Under his superintendency the centralizing trend of his three predecessors was reversed.

Mr Berry chose to decentralize the school district. A principal reported, 'I think the basic concept of decentralization as it exists in this district was the brain child of Berry'. One parent said, 'I would say that would have to be Mr Berry'. Another parent said, 'The key person was the guy who got hold of it and ran with it, Mr Berry'. Further evidence of this conscious choice and unconscious wish are the following examples from the data.

There can be no doubt that there were influences present in the culture. But again, as far as we know, why was it that only in Riverside did decentralization to this extent take place? Some informants have strong arguments to make about its origins. One administrator says, 'as far as power goes, if there was someone forcing Ray, I did not know about it. If there was a board member, I was unaware of that'. When asked if there had been community or parental pressures, he replied, 'I don't think that RUSD has ever been scourged by parents. The parents here had more confidence than do parents nationally'. Another officer said, 'It came from Berry. I think it was within him. I don't remember any outside pressure'. A principal said, 'Ray was a dynamic person that had some fantastic ideas and had some beliefs and philosophies that really differed from the general run'. Another principal summed up what all other informants were saying when they used words like his head, brain child, and beliefs and philosophies, by saying, 'Mr Berry was ahead of his time. He had a dream and it was far beyond anything he proposed'.

Another indication of this intention to improve the school district is that decentralization in Riverside was conducted within a 'framework'. Informants reported and understood that decentralization is a process which pulses, recedes, expands and draws back. There was always leadership which took the ultimate responsibility for guiding the entire organization.

Consistent with Ivey's (1969) view that intentional individuals act to respond to changing environmental situations, Mr Berry was described as an 'orchestrator and labeler: taking what can be gotten in the way of action and shaping it — generally after the fact — into lasting commitment to a new direction. In short, he makes meaning' (Peters and Waterman, 1982, p. 75). A description from the data shows that the superintendent could bang his fist on a table on occasion and the central office could say, 'Everyone could not always have their way but the disagreement was with mutual respect knowing the *intent* was there'.

The superintendent's intentions became obvious as each group experienced the changes that were being instituted in the organization. When one enters the field and asks school people to describe the district before 1960, they talk about a system which was academically strong, looked up to throughout the state as a beacon, and run by leaders who instituted new programs before they were state mandated and state funded.

Majority parents' comments indicated they had approved of the system generally. Black parents said it was an unfair system. It was an institution where the children on the hill got everything, and the children who were not from the high-rent district received hardly anything. Mexican-American parents tell about having their own school in one part of town which had had one principal for 41 years. That principal himself said that the school hierarchy felt that he would not fit anywhere else in the system (*Press Enterprise 1960–1978*, 1960–1978, 1977, pp. 1, 6). Mexican-American parents said they were isolated and without a voice in their part of the community. One of the parents said that some of their children were not developing as they should at that time. These comments are similar to those documented in the Fleischman Report describing two separate and unequal school systems in the United States (LaNoue and Smith, 1973). Not all children were included in the school organization.

Riverside did not differ in its degree of failure and success nor its structure. Its bureaucratic nature reinforced the inequalities perceived by the minority parents. For example, in describing books, the parents claimed, 'Books were ten years old in their schools. When the books got too old and too dirty for the White students, they gave them to us' (*Press Enterprise 1960–1978*, 1960–1978, 1977, pp. 2, 6). They also pointed to other differences, 'Some schools have fine buildings and teachers with master's degrees and others do not. The RUSD district business manager described the system at that time as "VEEEEEEEERY" unequal in the amount of monies distributed to the schools. And as usual, children from more favored families had more opportunities, and had read books most of their lives and had all kinds of experiences leading to better performance on tests'.

Parents and school personnel knew where the good schools were as well as where the poorer schools and teachers were. But the structure of bureaucracy serves to justify the retention of current practices in which the majority children generally tested higher than the minority children. An explanation is given in the following quote by a former district employee:

> Their parents are probably brighter. They are the ones who've done well in business and it takes brains to do that so the kids just naturally do better. Of course, there are exceptions, there are *many* and very outstanding exceptions. But the schools that served favored families always tested better. Children had had more advantages, they were probably brighter. And then the other schools had children who came and went...plus the fact that there were in those lower economic parts of town lots of people coming and going, many bilingual people. Children who had never read English, much less hadn't even spoken it, you know. So naturally they are not dumber, necessarily, but they don't do well on tests.

Another practice which perpetuated inequality was to exclude Blacks such as in the following example. A Black parent was made to feel 'Well, what are you doing here? You don't have an appointment and well, what do you want? We are the experts and you don't know... You are the dummy'. Riverside was probably no worse than most other systems in the United States — then something happened.

There were changes, and people describe and give examples of these changes. They relate to these changes first in personal terms, how they were affected, and then how the schools were affected. The description is one in which the organization changed from a centralized to a decentralized one and from one of exclusion to one of inclusion, because the superintendent intended that it be so.

A Parent Describes the Changes

One parent described what had happened to the schools by saying the district became open, non-uniform, non-structured, and ungraded. As a consequence, two of the family's children were placed in private schools to escape the district's public schools where the family felt 'the children were bumbling around in classrooms'. She knew one teacher who quit because 'she was a teacher and not a babysitter!' This parent was what others were to characterize as a 'hill parent' and had had no basic complaints about the schools before the changes and had liked them when they were structured. She asked, 'What would my family be like if we lived in circles (referring to Mr Berry's form of consensus decision-making)? We would really be in a circle!' This parent saw the cost of schools go up and, in her opinion, the scores go down. The change in the system had meant personal expense to put children in private schools and an active effort to present evidence to the board of education that the schools needed to go back to the basics. This family was concerned that the schools stay as they were and not become 'open and democratic'. 'Open' to them was a negative in two ways and decentralization represented open. First, it seemed their basic fear was the removal of God from the center of the school which was represented to them by the removal of structure around absolutes and allowing freedom and autonomy. The family saw the autonomy which developed in the Riverside schools as a type of self-love which is contained in and espoused as humanism. One tenet of humanism is decentralization of organizations and autonomy for the individual. Therefore, the introduction of autonomy and decentralization into the schools, for this family, was the introduction of humanism and the removal of God.

'Open' also seemed to be a negative to the family in that they equated decentralization and desegregation, and they saw the system as opened by an artificiality. They deeply believed that the artificiality of busing and integration had taken a humility away from a fine people:

> I think all people need humility. Humility means you can be teachable, understanding, have manners, and be kind. To be humble you don't have to be way down but true humility is greatness. What we have done, I think, is that because we have forced everything we have taken a humility away from a people that were marvelous.

The family said it did not look at the color of the skin but at the soul. Black families in their neighborhood were seen as good people and they spoke to each other when they saw each other. The mother had had a paper route for training her children and for the extra income for years. She said children of all colors had worked together happily out in the garage where they folded and prepared the papers. This parent represented those who were for equal education and equal schools and equal teachers but not on the grounds of what they termed artificiality. They felt a 'ride in the sunshine would not educate anyone'. These were active concerned parents who held firm beliefs and knew that other persons had opposite views which were held just as firmly. In having the pro and con about decentralization, both parents had taken their causes to the public forum. They felt the hearing they had received had not been objective but believed that their views were now vindicated.

Changes in the structure for this family meant that teachers conducted open classrooms in which children were expected to make decisions. The economic, religious and social changes which were taking place were viewed as negative. Other families interviewed saw the changes as positive ones for them personally, for their children, and for the schools.

A Teacher Describes the Changes

One teacher described her experience by saying that she first became aware of decentralization when her school took the leadership in looking at curriculum and budget. The assignment included preparing a budgetary and programmatic proposal for their particular children. The staff at her school started meetings, decision-making, and other innovations to make their school better. From these beginnings a new course of study which was originally called *A Third Culture* was developed. The new curriculum eventually was adopted throughout the district under the name *Man, A Course of Study*. The excitement for this teacher was that it was piloted at her school. She explained, 'It was most exciting because we could be creative, and it was exciting because through it we were able to do reading, writing, and math'. For example, her class chose to do a unit on Greece, complete with Parthenon Pillars and all, containing other curriculum areas besides Greece. The unit allowed the teacher to get into the affective domain of sensitivity in relationships, and cultural differences and similarities. The teacher's memories cherish the positive experiences of those days which included wonderful, supportive and willing parents who visited the school site and classroom.

This same teacher related how excitement filled the entire school and how the teachers didn't know what it was to 'go home before dark as an everyday thing!'. For her it was the best three or four rich years of what she felt education should be. She believed in the basics and the essential types of things that children must have, but teaching under decentralization was different. The teacher's impression was that such a development in one school could *never* have happened without decentralization and its basic philosophy of, 'We are going to put monies into schools — allow and support these teachers to do their creative kinds of things'. The creative experience she had was very rewarding.

Upon moving to another school where decentralization was handled differently, she found that the effects of decentralization depended on the leadership at each school, how school personnel took hold of the situation, and how much they really wanted to be involved and how much autonomy and authority teachers were allowed. At the second school there was 'no drive to really go for it'. The school did not innovate or include parents. Instead, the money was used to buy more material and to replace obsolete equipment that was not directly involved with curriculum.

In comparing the two schools, she thought the first school had accepted the primary goal of decentralization, which was to educate children and produce happy children: 'Not just to educate them to produce high test scores and attend Harvard or other comparable institutions, but to be happy and successful in other ways as well'. She saw the difference in commitment to that goal in the second school when she looked up at 5pm and thought in disbelief, 'Everyone's gone and it's not dark yet, and it is so lonely!'

The preceding interview data show the changes for the teacher were an increased involvement (she and the staff made decisions), a changed commitment (she and the staff worked long hours), and a changed relationship with parents (parents came to the classrooms and involved themselves in the planning). These changes resulted in high-er morale (excitement) and in greater productivity (production of a new curriculum).

A Principal Describes the Changes

Principals describe the changes as those from 'having to call the central office about every little thing and not being able to make any decisions that were not by policy or by the book to being allowed to do my own thing whatever that was'. It was not without rules, however. There were state and education code rules, but within a framework he felt free to work with parents, children and teachers to build a pro-gram for his school 'because each individual school had a unique set of problems and the ethnicity and economic makeup in each school was so different that they could not be treated the same'.

One change for this principal was that he became personally accountable. He and the staff had the budget and program to prepare for their school. This budget control was a startling change from the past when he had not been able to transfer from magazines to supplies and had been instructed how to spend each dollar. The results for him were freedom, autonomy, and accountability. He did not mind being accountable as long as he could use his own ideas and creativity to solve problems. He viewed himself confidently because he had respect from the teachers who knew he could make decisions and help them with their problems, and he did not have to take everything downtown.

This principal was aware that he had to demonstrate good judgment because of that added responsibility and authority. He said:

> You had to have really good judgment. Without that, with that kind of freedom and flexibility you could go out there and do some things that were really off the wall and get yourself in all kinds of trouble by doing things that really do not make sense. So it's easy to get into trouble if you're a screwball kind of administrator and you have all of this flexibility.

> Some principals that were not very creative, resourceful, or good managers really hurt.

After leaving Riverside to take a higher salaried position elsewhere he was disappointed in the district where he could not make the smallest decision. He now had responsibility without authority.

A District Officer Describes the Changes

Each level in the organization had equally dramatic stories to relate about changes which took place in the district and in themselves. One district officer described the change for him as a change of mind. At first he did not agree with the consensus style of decision-making which the leader was trying to develop in the central office and in the district as a whole. Although it was an intriguing concept, it seemed to him that it was inefficient and time consuming and that a lot more could be done more rapidly under a benevolent dictatorship which was the former work style of the office. He had to be convinced that the changes resulted in improvement.

The major changes which took place were a change in attitude and emphasis on the youngster. School districts are normally concerned about children, but decentralization in Riverside altered the concern of the central office from managing the district and, only incidentally, operating the schools, to the new concern described thus:

> Certainly we have a business office and we have a support services office. But the only reason that they're in the district is to provide service to schools. And that is the attitude that came about; no longer was someone up here saying, 'You do this'. It changed so that what was said instead was, 'How can we help you do whatever it is we're jointly planning to do?'

What this central office administrator described was a revolutionary change in his thinking about how you work with people in an organization; the evolutionary change in the district was from one of ordering to that of listening; and from a primary interest in managing adults to an interest expressed in joint planning for children.

The changes which each participant described — parent, teacher, principal, and central office administrator — occurred in two areas: (1) in the personal, and (2) in the organization with significance for both. Examples were then presented regarding the ways in which schools were affected. The changes were from a centralized, bureaucratic, hierarchical school district to a decentralized school district. The changed social relationships included consensus decision-making. Principals had the opportunity to lead; teachers became more committed; parents were given the opportunity to be active in school affairs; and children became the focus of the district. Negative criticisms were from those who, in disapproving of the changes, equated decentralization with no control. They considered non-graded classrooms, lack of structure, and requiring children to make decisions as evidence of individual autonomy which they equated with self love. Other criticisms were voiced by principals who were made uneasy when power was transferred from the superintendent to them. Each of the criticisms adds evidence to our argument that the profound changes described were from centralization to decentralization.

We have looked at what persons described as changes in the district from 1960 to 1978. Next we present data from an analysis of the Riverside Unified School District Directory from 1959 through 1979 which also demonstrates changes from bureaucracy to decentralization.

Table 2 Positions in the Riverside Unified School District Central Office, 1959–60 to 1978–79.

Year	Number of Positions
1959–60	30
1960–61	37
1961–62	38
1962–63	48
1963–64	42
1964–65	41
1965–66	64 (high)
1966–67	30
1967–68	25 (low)
1968–69	41
1969–70	34
1970–71	28
1971–72	27
1972–73	27
1973–74	26
1974–75	25 (low)
1975–76	26
1976–77	27
1977–78	30
1978–79	31

Source: *Public Schools Directory for Riverside County* (1959–1978), California.

Organizational Evidence

Examination of the public schools' directory for the Riverside Unified School District for the period 1960 through 1978 showed that the district was once highly centralized and indicated that: (1) the number of positions, (2) the kinds of positions, (3) the titles of positions, and (4) the emphasis of the central office had changed. These changes resulted in less centralization in the district. Through the 19 years there was a total of 142 position titles and 128 persons held those titles. The number of positions peaked at 64 in 1965 and hit a low of 25 in 1974. Numbers per year are shown in Table 2 and Figure 1.

The high of 64 positions in 1965 can be partially explained by the addition of 14 nurses and 4 speech therapists. Those 18 positions were all removed from the roster the following year. The nine school years from 1959 through 1968 averaged 39 persons in the office, while the years 1968 through 1978 averaged 29. The school population figures are shown in Table 3.

The difference in school population from the highest year to the lowest was 3869 students. The difference in the district office from the highest to the lowest is 39 positions (or 21 positions if we adjust for the 18 nurses and speech therapists' one-time listing). There were far

Figure 1   Use of Word Comparisons in Central Office — 1960–1978.

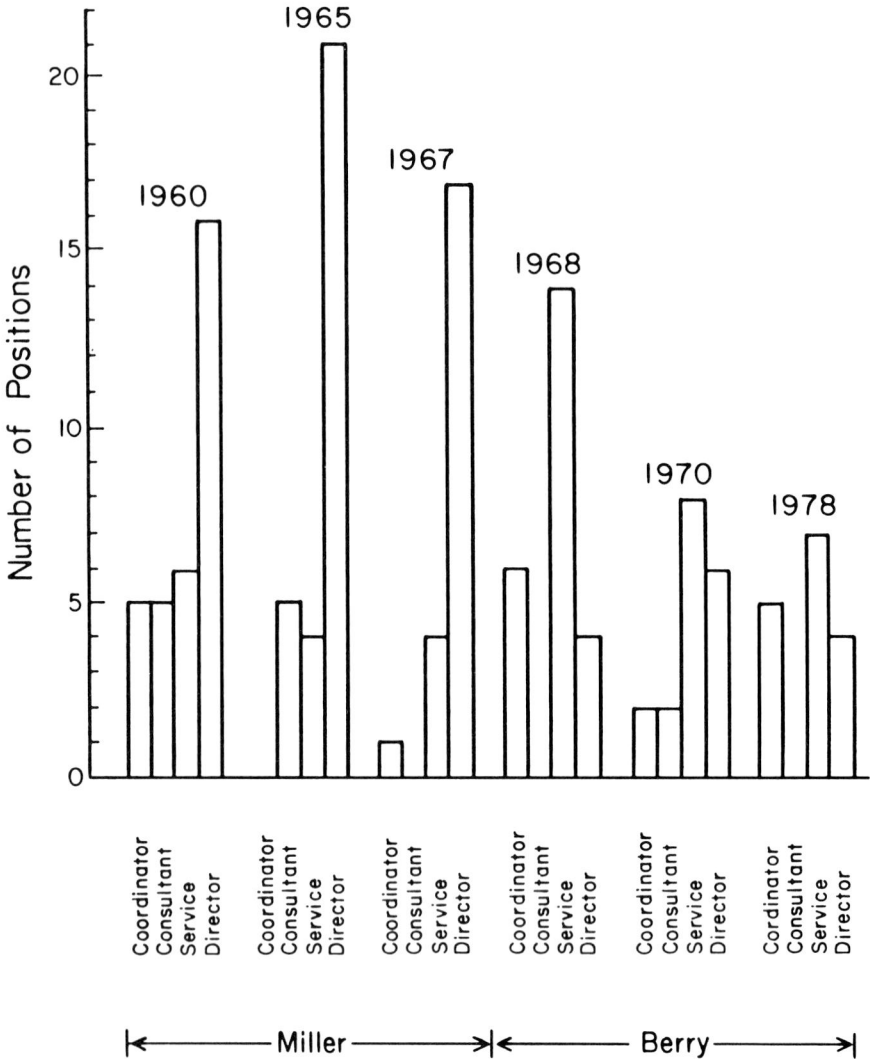

Source: 'Public School Directory of Riverside County, California, 1960–1978'.

fewer persons in the district office during the Berry years (1968–1978) even though in three of those years the student population nearly equaled or exceeded that of the previous superintendent's tenure. The difference cannot be explained as a function of a building program during the earlier time since there are only two positions listed in the directory for school house planning, and one of those continued into the Berry years. Neither can the difference be explained by fewer dollars (see Table 4).

Table 3   Pupils in Riverside Unified School District and Number of Central Office Positions, 1963–1978

Year	School population	Number of positions
1963	24,044	42
1964	24,928	41
1965	25,685	64
1966	26,021	30
1967	26,366	25
1968	26,912	41
1969	27,126	34
1970	26,794	28
1971	26,192	27
1972	25,376	26
1973	24,773	26
1974	24,308	25
1975	26,138	26
1976	24,182	27
1977	23,663	30
1978	23,257	31

Source: Figures of Population from Riverside County Schools Office, 1983. *Public Schools Directory for Riverside County*, California, 1963 to 1978.

Table 4   Riverside Unified School District Budget and Average Daily Attendance, 1963–64 to 1978–79

Year	Budget	Average Daily Attendance
1963–64	11,690,580	24,455
1964–65	12,754,900	25,070
1965–66	14,244,162	26,033
1966–67	14,416,640	26,763
1967–68	15,686,992	28,461
1968–69	18,533,114	27,785
1969–70	19,479,154	27,904
1970–71	22,098,377	26,685
1971–72	21,917,221	26,780
1972–73	24,924,226	25,293
1973–74	25,503,001	25,119
1975–76	29,929,682	25,456
1976–77	31,994,002	25,571
1977–78	38,735,567	24,192
1978–79	37,060,777	—

Source: Courtesy of Riverside Unified School District Office, 1983.

There were changes in the kinds of positions in the district office, too. In 1961–62, (an average year during the previous superintendency), there were 38 positions in the district office, and the following positions existed to oversee directly subject matter areas in the schools: consultant of physical education, coordinator of industrial arts, coordinator of science, coordinator of mathematics, coordinator

Table 5   *Riverside Unified School District Central Office Position Titles, 1960—1975.*

Word	1960	1965	1967	1968	1970	1975
Director or assistant director	16	21	15	5	6	4
Consultant	5	5	0	11	2	0
Service(s)	6	4	4	14	9	8
Coordinator	5	0	1	6	2	5
Instruction	0	1	1	7	4	2
Total positions listed in directory	37	64	25	41	28	26

Source: *Public School Directory for Riverside County*, California, 1960—1975.

of art education, director of dental health education, and consultant of elementary education.

In 1971—72, an average Berry year with 27 in the district office, all the above positions were gone. The emphasis had become one of directing broader programs rather than tight control of all subject matter. New job titles reflected this broader view: consultant, EMR/gifted; consultant, educational handicapped; director, occupational education; director, special programs, elementary curriculum. To the elimination of jobs and an emphasis on a different type of job overseeing broad programs was added a change in the words used in job titles in the district office. For instance, note the number of times the word 'director' appeared in titles prior to the year 1968, in Table 5.

In the first Berry year, 1968, the heavy use of the word 'director', which implies more authority and more control, was largely replaced by use of the word 'consultant' or 'coordinator', which implies a consenting peer relationship. It is particularly instructive to compare the use of words in job titles for 1967—1968, the last Miller year as superintendent, with 1968—1969, the first Berry year as superintendent. (See Table 5 and Figure 2.)

From 1966 to 1968 there were six changes from the word 'director' to the words 'consultant' or 'coordinator' in titles; five instances of 'director' titles which were completely eliminated, three title changes from 'director' to 'supervisor', and two 'director' titles which had the word 'services' added to them (see Table 6).

There were other word changes (additions) in the directory. The principal aim of a school district is that of instruction yet the word was not emphasized in the district office titles until 1968. The word 'instruction' was not used in 1960, was used only once in 1965 and 1967 and seven times in 1968. The importance given to instruction becomes apparent in comparing the number of positions listed under instruction with the number of positions in the three other areas. The

Figure 2   Central Office Titles, 1968–1978.

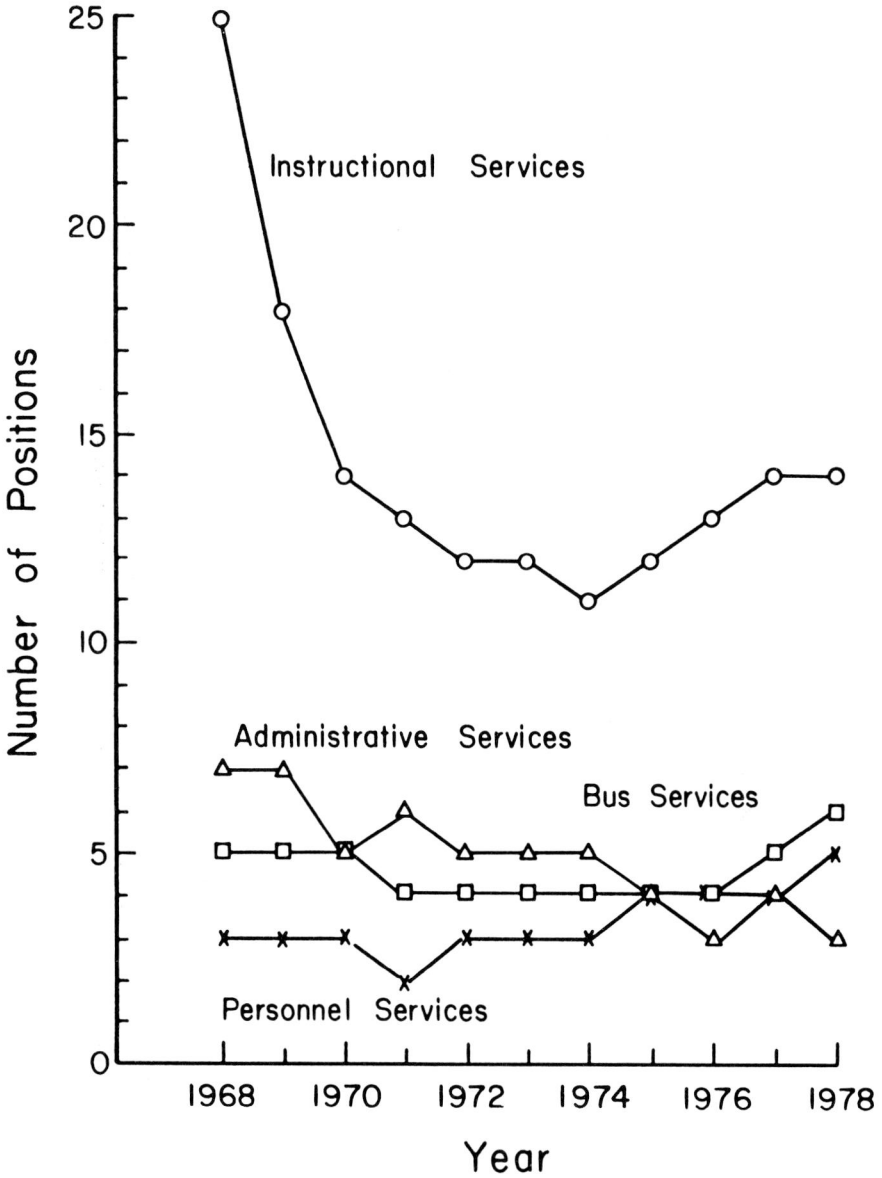

Source: 'Public School Directory of Riverside County, California, 1960–1978'.

Table 6   Changes in Titles of Positions, 1966–1968 and Elimination of Titles, 1966–1968.

1966	Titles, 1967	1968
Director, secondary education	Director, secondary education	Eliminated
Director, special education	Director, psychological services, special education	Consultant, psychological services
Director of audio-visual education	Director, audio-visual education	Coordinator, instructional media services
Director of instruction and curriculum development	Director of instruction and curriculum development	Director, instruction and curriculum
Director of reading	Director of reading federal	Consultant reading
Director of elementary education	Director, elementary education	Eliminated
Director of guidance	Eliminated	Eliminated
Director, summer school	Director, inservice and summer school	Eliminated
Director, inservice secondary interns	Director, inservice summer school	Eliminated
Director, personnel	Director, personnel	Assistant superintendent, personnel
Director, library services	Director, library services	Coordinator, library services
Director, child welfare and attendance	Field supervisor, child welfare and attendance	Supervisor, child welfare
Supervisor transportation	Director, transportation	Supervisor, Transportation
Director, psychological services	Director, psychological services, special education and counseling and guidance	Director, instruction, pupil personnel services
Consultant, social studies	—	Specialist, instruction, social studies
Controller	Business manager	Business manager
Director, research and testing	Director, research and testing	Director, research testing
Director, maintenance	Director, maintenance	Coordinator, maintenance and operations
Director, cafeterias	Director, cafeterias	Coordinator, cafeteria services
Director, health services	Field supervisor, health services	Supervisor, health services
Coordinator, gifted program	Coordinator, gifted program	Consultant, gifted
Director, intergroup relations	Director, intergroup relations	Supervisor, intergroup education

Source: *Public Schools Directory for Riverside County*, California 1966–1968.

number is always two to three times larger than the number employed in personnel, business office, or administration.

Also, we find that there was a change of emphasis in the directory to a focus on service. Beginning in the year 1968 all positions in the directory were placed under one of four logos, each of which ended in the word 'services'. No such headings were used earlier.

This analysis of the district directory for a period covering 19 years has shown that the number of persons in the central office declined from an average of 39 to 29; the titles generally were changed from director to consultant; the types of positions became less controlling ones; the word 'instruction' began to be emphasized; the instructional component of the central office was always at least double the size of any other division; and that there was a new and clear emphasis on services being rendered by the central office to the rest of the district. From these data we are able to form a picture of the centralized district inherited in 1968 by Berry and the changes he brought about.

Evidence of Intentionality

The data demonstrate conclusively that the process of decentralization in Riverside was triggered by forces in the organization. One way to discover if decentralization was an internal or an external process was to look at when it began in Riverside. Informants give different dates for its beginnings depending upon how close they were to that beginning. If the process had been ordered by a central office directive on a specific date, the questions would not arise. It was not so ordered. The process itself was conducted in an evolutionary manner. Therefore, according to a principal, 'the advent was a slow and tedious process'.

'Decentralization first began to be used as a concept and a goal in the Riverside Unified School District in 1962. The process began to take on form and substance as an official direction and goal with a report to the Board of Education in January 1966' (Berry, 1971, p. 10). Though bureaucracy is noted for its love of, and use of paper, nothing was written officially about the process until nine years after its introduction as a concept and goal. One informant said this was the case because Berry believed, 'You are all creative and if we put this in writing, the creation will stop and people will see it as finished and set in concrete, and I want us to continue to create'. One principal said:

> It was an evolution in the system. A very massive effort to remodel the organization 1959–1962. When I went out as a principal in 1965 I was one of the first to implement it... Ray deliberately refused to put anything in writing.

In this statement one is able to see an organizational remodeling before 1962 (Berry began in the district in 1960 as personnel officer), and implementation in the field in 1965. Two other persons gave 1962 as the start-up date. One central office person said, 'There were some initial steps toward decentralization while Ray was Associate Superintendent (1962)'. Another confirmed the same date by saying, 'Ray began working on a decentralization process during his tenure as Associate Superintendent (1962)'. One statement by a principal is important to our contention that Riverside's was an internally and intentionally-triggered decentralization process:

> This was the Civil Rights era and I really believe that one of the reasons why we went through the Civil Rights thing in Riverside Unified School District *almost* unscathed was because we had developed a decentralized decision-making process that involved parents already... Decentralization came first in this district before desegregation.

The date was important to establish in order to be certain that desegregation of the district in October 1965 was not *the* trigger. Several informants remembered the two as somehow connected. And they were. Desegregation and integration grew out of the same philosophy and effort which put decentralization in the district. Berry stated that, 'the two — decentralization and desegregation — are of a piece and that you had to have them both in place or else you would have an explosion and that's what had happened in so many districts and is happening now'. In Riverside the date is clear. Decentralization came first in 1962. Desegregation was accomplished three years later and was one outgrowth, at least partially, of decentralization. 'Berry used desegregation to further what he already believed in... It was a convenience', one teacher said.

The literature cites ten influences in the nation which triggered decentralization in public schools. There were cultural influences throughout the nation pressuring schools toward decentralization. However, one must note that decentralization did not become an issue in New York City until 1966, and its decentralization law was not passed until 1969 (LaNoue and Smith, 1973). That law was the nation's first. By 1969 the concept and the goal had been in place for seven years in Riverside. And it had been an official goal for three years and in place in at least one school since 1965. The literature shows that even after laws were passed in New York and in Michigan, there was little change in the decision-making levels in the schools (Fantini and Gittell, 1973). Professionals resisted effectively. We have learned that was not the case in Riverside. The process moved decision-making out to all levels of the district and the central office became a service, not a 'command structure'. There is no evidence in the literature that this occurred in any other school district nationwide at this early date. In fact, in 1984 one informant declared that as a consultant to school districts, in all of the counties in California and 38 states, he had not found the 'type and the degree of decentralization in any other district that he had experienced in Riverside Unified School District, not with the spirit, not with the commitment'.

Cultural factors influencing decentralization in Riverside are cited in the data. We have already seen that Riverside began decentralization before it desegregated, but the effort is remembered together by some and in reverse by others. A board member says:

> I really associated it as beginning to emerge out of the integration effort, so I don't know exactly but I think our decentralization came about because of a need more than anything. As we started moving into the integration program we found we had to have very good people out in the field and we had to have people who could think, who could make decisions. You weren't going to be able to tell them. There were 10,000 different ways in which we could get into trouble on this because there were decisions that had to be made at every level and everybody was going to be involved. Everybody had to be assuming a lot of responsibility of working together. It's one of those things where you have to keep winning every single little decision along the way to have a successful program and there are so many opportunities to falter.

The two are probably remembered in reverse by the board member because, while desegregation was quick and dramatic (seven weeks), decentralization from its beginning in 1962 'took all of Mr Berry's tenure in the district and it was a step-by-step process', and was not officially presented as a goal to be endorsed and voted on by the board until after desegregation in 1966. It is also possible that a skilled organizational leader would use desegregation to increase decentralization if decentralization had been an important earlier key belief and philosophy for him. One should note here that it appeared that a need for economy in the district was also seen as an opportunity to press forward on decentralization of the central office. Berry said, 'When you have these concepts in mind, when the opportunity comes along, you take advantage of it'. Though he was speaking of changing personnel, he certainly would take advantage in a most positive way of something as helpful as desegregation to help fulfill a primary goal of his.

A central office administrator said, 'I don't know if we were the very first. Decentralization became a very common word in all kinds of state-wide meetings'. Federal programs such as the Elementary and Secondary Education Act (ESEA) and Early Childhood Education (ECE) started about the same time as decentralization in Riverside but as one principal noted they 'meshed right with our philosophies that were being formed. There was a feeling in the district that society was passing so quickly, that we needed to open things up so that we had more input from parents and teachers and from kids themselves'. There were other such cultural influences as the non-grading, team-teaching and multi-age grouping. Those kinds of things became very popular. And finally one central office administrator said, 'I think there was a climate in the community as well as in the country which had begun to encourage this, but within our own school setting I really have to credit mostly Ray Berry'.

All informants were in total agreement that the idea, the impetus to implementation, and in a few cases the imposition of decentralization came from Ray Berry's head. The business manager said, 'I recall when Ray started this 'thing'. Another person said, 'He used the term before I knew what it meant'. In the Riverside Unified School District, decentralization and Ray Berry are said in the same breath. One administrator said, 'Ray Berry was probably the sole supporter and sole innovator in the decentralization process early on'. A central office person said:

> Decentralization happened here and he was the one who started it. There has never been any question in my mind. I'm almost positive the board had to be taught — or most of them probably. But he was certainly the one who brought it to my attention.

A principal reported, 'I think the basic concept of decentralization as it exists in this district was the brain child of Berry'. All parents agreed.

Chapter 3

The Process of Decentralization

Decentralization in Several Stages

Decentralization went through seven stages. The first stage began when Mr Berry arrived in Riverside as Personnel Director in 1960 and became acquainted with the structure and the people in the district. He first presented his ideas and concepts to Superintendent Miller, who encouraged him during this first crucial stage. At this time he experimented with projects in some schools, created committees and searched for ways to increase organizational participation.

In the second stage Berry 'got hold of the organization'. He became superintendent upon the retirement of Miller and he presented, communicated, explained and articulated bits and pieces of his values and concepts to the board. He used his projects and committee structures as support. His outstanding board supported him in this second stage.

In the next two stages he concentrated on the central office. He had been 'tinkering' there during the first two stages and he would continue to rearrange central office staff almost every year, but major moves were made between 1968 and 1969.

In the fifth stage he moved further out in the organization, selecting school sites to work closely with some particularly interested principals and teachers. Stage six was to get them all together, central office, principals, and teachers, to form clusters. During the cluster stage he provided opportunities for school personnel to talk and communicate with each other and involved the teachers in relating as peers to principals and central office personnel. The sixth, or cluster stage was called by one principal 'the very heart of the implementation of the dream'. In stage seven as superintendent he was gently recentralizing to maintain stability. It took 18 years to complete the reorganization.

None of these stages is as clean as a surgeon's cut. Berry was working out in the schools with some principals and teachers even before he made major moves in the central office. He was also talking to the board at the same time as he talked to Miller. If one were to have had a bird's eye view of what was happening in the district during these years, one really would have seen Berry moving on all levels to some extent at the same time but with different intensities and with different programs and strategies. He used every opportunity which fell into his path and he created opportunities. He knew broadly what he wanted in the way of a finished picture, but he did not know exactly how to paint it.

Decentralization is a Process

Decentralization is a series of processes which, when added together, equal a whole. For example there is the process by which the leader reaches the point of risking to change the organization. Then there are the actual steps taken to implement the changes. Next there are varying changes in each level and in each person which can be characterized as processes which trigger secondary waves or ripples. And as we will find later, decentralization is not a 'once learned, always in place' change process. Organizations are as dynamic as the human body. The individual members are in constant flux. The process must be repeated over and over and re-learned repeatedly if it is to stay viable. So when we speak of decentralization, we mean a continuing process on which Berry continued to work as long as he was with the district.

'The process was one where it was necessary to allow a lot of things to go on that normally wouldn't be acceptable. And it was risky and it was daring', stated a principal. 'The answer to "what happened?" is often arguable' (Sproull *et al.*, 1978, p. 10).

To understand decentralization in Riverside one must walk step by step through the process with the participants and the organizational leader. The one common word used in the data by nearly all the professional participants to describe the process was 'emerge'. Berry said, 'I'm not sure it was conceived as a *total* plan in great depth. I think it emerged from some of these feelings in bits and pieces'. A central office person said, 'It just grew or emerged, you might say'. Others said it was piece-meal, step-by-step, and grew through the years. The conception, implementation, and results all emerged. The process went through stages which were easily seen after the fact, but which were buried in the head of one man before they materialized.

Step 1: Conceptualization of a Dream

The process begins with an organizational leader who had a dream. The dream is attributed to him by those who took part in the process, or who were close to the process. In fact, they say, 'He had a dream so big that he only proposed it in little chunks'. He says only that he felt 'almost instinctively driven'. With that inner drive he set out to do the things he did, decentralizing a school district. He said, 'it just seemed natural that I would do those things'.

Burns (1978) wrote of leaders having a vision and of being compelled to complete the vision they see. One principal said, 'Ray is a man of vision'. Brittenham (1980) found that leaders must: (1) be some stages ahead of their followers; (2) have a well-developed philosophy to sustain themselves when problems arise between intention and actualization, and (3) demonstrate integrity. He says, 'Leaders in school districts...completed the philosophy and vision...before involving members' (p. 152). Berry was said to be far ahead of his time, and 'to be able to see down the road many years'.

He thought about his dream a great deal. He was known as an internal and a deliberate man. A principal said, 'He is not the kind of man who has a dream one night and comes down to the office the next day to put it into place'. Principals speak about his being an avid reader and about his mulling over ideas. Berry explained, 'I thought about this a good deal, or I searched for an answer'.

An indication of the intense and continuing thought which the process required is shown in Mr Berry saying, 'it just consisted of hundreds of things. I took hold of an organization and looking back on it 15 years later, it was not the same organization at all'. Any process involving hundreds of pieces was an intellectual undertaking. The statement, 'I took hold of an organization' demonstrates that the process was one of active intent initiated by someone with a purpose. So the first step was the presence of an organizational leader with a vision who had hold of an organization. After thinking about it a great deal, he took steps to actualize his vision.

Step 2: Communication of Concepts

As stated earlier, the discussions about decentralization began as early as 1962. It was necessary for Berry to share his ideas with Superintendent Miller and the board president so that they would understand and be supportive of what Berry characterized as a radical departure from what the district had been involved in for over 30 years.

The president of the board of education, the superintendent, and Berry had lunch together about once a week, usually on Mondays. The president remembers personally discussing with Berry the directions the district ought to be going. He recalls discussion of decentralization and the broad policy matters associated with it. It was from these discussions that both the president of the board and Miller first learned about decentralization. The president said, 'I provided input but decentralization was not something I started. It was something that as I recall was really started by Ray Berry'. Berry's sharing of ideas struck no one as dream-sharing since he conversed easily and was a good listener.

What did Berry share at those luncheons? Berry had one driving overall vision: to educate each child effectively. He realized it would take every person in the district and the support of the community to accomplish this task. His deeply-held convictions and values drove his vision:

> The root of all this for me was that we were finally beginning to try to do what we said in this country we'd always do, which is to educate every child effectively. We never had. We hadn't really come very close. And the more I worked with it and thought about it the more I felt that it was such a huge task that it simply had to be done this other way, through decentralization. We couldn't do it, we had not that much control over the lives of human beings to force people to that kind of performance. They don't respond that way anyway as we know, so we had to reach for it this other way.

The above is illustrative of Berry's thinking, struggling with actualizing the mission of the school. He had a second belief that enabled him to proceed. Berry believed in the power and goodness of people. He said, 'I really believe and I still strongly believe that the roots of our country and our whole concept of democracy are based on the strength and power of individuals'.

He saw the role of leadership in a school district as releasing that power. He put it this way,

> In order to release that power you have to place trust and faith and confidence in those people and let them know it and give them room to exert their power and their capability and their creativity within some defined framework. And I see that as the role of leadership in a district.

Berry would have shared this thinking in the form of what he called concepts:

1 A democracy must educate children effectively. We never had.
2 Each person has power to give to the process if it can be released.
3 The structure and/or the way persons work together in a bureaucracy have to change so people will be free to use and to give of their power.
4 Structure can be changed by changing thinking, behavior, and relationships at all levels.
5 The redirection and program that a school needs must have parental and community understanding and the only way for the community to understand is through participation.
6 The role of leadership must change.

These child centered concepts were not all stated or shared at one time with the luncheon group. They emerged over the years as the men talked to each other. From these shared discussions Riverside began the trip down a new and virtually unexplored organizational road which was later (in 1971) named a *new direction*.

The ideas of decentralization, once put into words in discussions, were mutually supported by the board of education and Superintendent Miller. A principal said, 'Berry carried the board with him on all these matters 100 per cent'. Berry stated:

> Mr Miller sensed that this was right. He cared very much about children and people. I think he recognized it and I believe that he recognized that the *outstanding* board of education, that these people who were just top flight leaders in their own right, were also going along with this and most of them were pretty conservative community leaders. And for them to buy into some of these concepts of decentralization and desegregation was a strong signal to a man like Bruce Miller.

A district officer said that 'Miller was a fine man who subscribed to whatever was going on in education at that time'. It must be noted that decentralization was not going on in education in 1962 when the process started. Miller, however, who did not practice and live it, was willing to let it happen. Berry said Miller did not equivocate. He placed the whole weight and the authority of his office behind the idea of decentralization and said, 'that's what I believe in and here we go'.

Berry, who calls himself the 'pilot and legman' of decentralization, admired Miller because 'he was so willing to give me so much responsibility and authority and support'. When asked, 'Why?' Berry said he did not know, 'Because I don't think we ever discussed it that way'.

In retrospect, we can be certain only that there was no recorded resistance on the part of Miller — a most traditional superintendent — to a radically new direction for the district. It seems paradoxical. Perhaps he could support it because it was slow, evolutionary, and did not change the system overnight. Perhaps Miller, as the board president remembers, did not really ever vote on the whole thing because there never was a whole thing. The board president stated:

> It's the kind of thing that didn't happen in one day and we say, 'Oh, well, today we are going to change the whole system around'. So, it's the kind of thing that you really don't ever see in a context where you would suddenly say, 'yes' or 'no', in terms of content unless you get down the way and then you see the results.

Decentralization, in concept and in practice began in the district in 1962. It took form and substance as an official district goal and direction with a report to the board in January 1966 (Berry, 1971, p. 10). Miller is first recorded as using the word in a board meeting when he introduced it in regard to a projected district deficit of $700,000. In making his recommendations he stated that, 'two basic considerations were used in arriving at the list of adjustments. One was a new direction involving the principles of decentralization and the other was the need to demonstrate economy' (Riverside Unified School District, 1960–1979, 2-27-1967).

Miller then called on Berry, as Associate Superintendent, to outline the ideas underlying the new direction as 'an integral part of the superintendent's recommendations'. Next he presented his preliminary recommendations for program adjustments which were considered necessary if a scheduled tax override failed. Thus, decentralization was first used in the recorded board minutes for the first time in connection with saving money. One month later the program reductions which included the principle of decentralization were approved. Two months later the tax override was passed. (Riverside Unified School District, 1960–1979, 5-29-67.)

The vote by the board approving the new direction and the principle of decentralization was ratification of a process which had been underway for four years. The board approved the principle without a list of specifics. This is probably why the board president could say that decentralization had never had a vote on its content.

Even though the tax override was passed and the budget continued to increase by as much as one to seven million dollars each year for the next seven years, the decentralization process continued. Its second mention in the board minutes was on December 15 1969, when a principal said that he spoke for many other schools in praising the success of the decentralizated budget. At that point board members expressed some concern:

> Board members commented on the possible dilemma in the setting of priorities, of the board's infringing on the decentralization process, and it was stated that the board should make the broad priorities and objectives to allow flexibility at the schools (Riverside Unified School District, 1960–1979, 12–15–69).

Mr Berry responded that 'the framework of education is rigid and the district had reorganized its structure to give it more flexibility'. The questioning of the board members almost four years after the board had been introduced to decentralization, and three years after it had become an official goal of the district, and two years after the board had voted to approve the principle of decentralization, indicates at the least that all members of the board were not yet clear how the process worked; or they were not clear exactly what their role in the process was; or they did not favor it completely. The concern of the board was most likely a real desire that they do nothing to hinder the process, for Berry and two principals stated that the board was 100 per cent behind the process.

One reason that understanding decentralization was problematic was because nothing was written down until 1971. When the decentralization process began no one knew what it entailed or how to accomplish it. There were no models or patterns in place. Berry said, 'This was long before or at least some time before the national pressures, laws, and the court cases were causing these things to happen. In our case we were somewhat ahead of it because we believed in improving our schools'. They were also ahead because Mr Berry in his association with the University of California, Los Angeles (UCLA) and Dr John Goodlad engaged in a special project. In 1965, Goodlad and his UCLA researchers began the League of Cooperating Schools to study how a principal becomes a 'change agent' in a school. One Riverside school was among the 18 schools chosen to participate. According to one principal UCLA was as interested in Riverside's effort to decentralize the whole district as Riverside was interested in UCLA's work in individual schools:

> The UCLA staff was interested in Riverside's direction of decentralization of the district. They felt that it hadn't been done before and they wanted to know whether it really was going to happen ... They were curious and watchful as to what was happening in Riverside.

By 1965, across the nation, efforts towards decentralization were encouraged, but no district in the United States had done so. But even in the face of pioneering, the board president 'remembers no opposition on the board to the concept'. What did people mean when they talked about decentralization? Though expressed in different ways, decentralization meant for most a change in the decision-making process. The president of the board said:

> Decentralization within the school system was thought of more in terms of change of command, as how much authority people had. But as part of that process ... there was very much to this effort to involve not just teachers and staff in the decision-making process but a lot of other people in the community, all kinds of people that also had an interest in our schools.

A principal said, 'We thought decentralization was allowing the decision to be made as close to the action as possible and that was basically with the teachers'. A parent said, 'I think it was his way of giving principals a chance to do what they wanted to do to express their own ideas and to use what they knew as principals'.

The president said that all kinds of people were to be involved. The principals said it was to allow teachers to make decisions, and a parent said it was to let principals make decisions. There was a general understanding that the process involved moving decision-making down the hierarchy all the way through and out to the community. How is that to be done in a bureaucracy which had been honed through three superintendents to have the professionals at the top make nearly all the decisions? Once the superintendent and board were supportive of the new direction, Mr Berry began the organizational changes at the central office level.

Step 3: Changes in the Central Office

Four fundamental changes took place in the central office: structure was changed by reducing personnel, behavior was changed by example, thinking was changed by instruction and communication, and relationships and attitudes changed. Berry realized that the central office was making most decisions for the district. If decisions were to be decentralized to other levels, the central office had to relinquish power. Homans' (1950) work demonstrates that when activities and interaction increase, positive feelings can increase in a group. Berry used increased interaction and activities as one way to change the central office. Or as Brittenham (1980) would say, he was several phases ahead of his followers! One administrator recalled:

> Berry alone in the central office favored the process when he began it. He was the sole supporter and the sole innovator in that decentralization process early on. He really had only one other person that was available to him and helped in that direction. That was one man we called the thinker, creator, at the central office level.

There was covert and overt resistance from central office holders to relinquish power. The specialists, directors and assistant superintendents had struggled hard to get to the central office. Their organizational preparation and experience had made them highly skilled professionals as well as loyal to their placement. Their work style was one of commanding and directing. Mr Berry wanted that changed to a work style of servicing schools. Berry said, 'We kept pounding away, almost by attitude and personal nature rather than constantly voicing it, that the central office is primarily a service organization, not a command structure'.

One administrator expressed it this way, 'All other administrators, except the thinker, looked at this direction with a jaundiced eye and with fear and trepidation and concern for their own positions'. The board president stated that, 'we changed a lot of people'. The changes came about in several ways. The structure was changed by reducing personnel. Prior to the time he became superintendent, Mr Berry worked to change some of the assignments of the specialists whose control of instructional programs was nearly total. Positions were eliminated through retirement. The director of elementary education retired and was not replaced. Some jobs such as the director of secondary education were phased out. One person was asked to leave because his ideas did not fit the new direction and another left because he related to people negatively. These actions removed one layer of hierarchy. Berry described the process:

> We can no longer have, or afford, in some instances, this much of a hierarchy — this much of a group of people at the central office level who directed programs. And as we moved through this we actually eliminated supervisors of physical education, fine arts, reading, and music. We just closed out the jobs and said, 'This is going to be it'.

The restructuring took place as the new ways of working with people and the sharing of decision-making increased. Berry explained, 'So when we began to open up the system through some of these changes it was almost automatic that some of the changes in personnel occurred. Because people who are uncomfortable with it began to find options for other choices'. Those who remained and those who were brought in faced an organization which was changing. Thus, there was not only a change in the managerial structure in the district, but also in the kind of people who fit it.

Step 4: Cadre Formation

As a result of central office changes Berry began to form what was called a 'cadre of managers' around him who complimented his strengths and weaknesses. Those persons eventually became emissaries of the program as they helped in the process as it evolved step-by-step. The persons who were left in the office and the new ones who were brought in learned to work and interact in a more personal manner. The group learned to work in an atmosphere of trust and informality. 'Other managers were quick to report an administrator who was being authoritarian or who was operating in an authoritarian manner', it was said by a district administrator.

Their commitments to, and their understandings of, the new direction varied. One method used to increase their understanding was by example. Some told of watching Berry use the process over and over as an example of how it would work. They were encouraged to participate and help use the process themselves to solve the real district problems. One said there was more of them being asked, 'Here's a problem. What do you think? What would you do with it? and, How would you implement that?' There was a constant tapping of brains and continual sharing of ideas.

Central office personnel had to become convinced that decentralized consensus decision-making worked. Not all of them were ever totally convinced. But those who remained in the 'cadre of managers' were largely those who understood and approved, at least in public, of the new direction. Others stayed, not because they were necessarily advocates of decentralization, but because they were excellent organizational men. One was asked if he were convinced decentralization was a positive way to manage an organization and he answered:

> I don't think it is a matter of being convinced. It is a matter of organizational discipline. If the man you work for is assigned the leadership — we are not running a democracy or a *laissez-faire* — if you are going to work there you had better see that his direction is implemented cheerfully.

There was one person who was described as the creator and thinker. One principal said that, 'Berry selected a gentleman who was what we called the creator or the thinker who would sit with him and discuss philosophies and directions and why nots and if you did this and so forth'. He was called a complex man by some and an enigma by others. It was said by a principal that:

> He was a very intelligent person, but he had a problem in projecting himself in large groups. He was not a vivacious man nor was he a personality plus, and you wouldn't know that he was there. And Mr Berry recognized this early on and recognized also the fact that he had a great deal of intelligence and also that he was very creative and widely read and had some ideas similar to Mr Berry's. He was used as sort of an assistant that was over to one side. People would look at him and say, 'What the hell does he do?' And nobody could understand it specifically unless you actually worked with him. He would always be the first one to write a speech and say, 'Here it is. Do what you want with it'. He was always one who could do the leg work and research and bring it to the fore and Ray used him well. And he was very happy in that position.

However, a central office participant said that, though the intellectual understood and wrote about decentralization, 'He did not support it and he spoke with a forked tongue'. It is not unusual for an intellectual, who constantly analyzes first one side and then the other and who is involved in the arguments on one side today and the other side tomorrow, to be seen as uncommitted to a cause. It was difficult to know just exactly what the role and impact of the thinker was. One person explained that he was the one who provided the analysis for the decentralization process. Another saw him as an academician, not as a practitioner. Perhaps that is why he was seen as not supporting decentralization. What is important is that there was a person recognized as an intellectual in the process and the organization acknowledged him. His ambigious role was described by one as, 'He was very happy in that position', and another as, 'He was not very happy'. The only agreement was that there was a creative thinker, an intellectual, and that he was deeply involved in the process. (See Masters, *et al.*, 1964, and Steinbrunner, 1974, for a discussion about scribblers and intellectuals respectively.) The role of this person is more fully covered in Chapter 5.

There is evidence in the data that persons who did not want the process to proceed and succeed and who were not committed to either decentralization or desegregation stayed in the central office. When the researcher asked if lack of commitment was a failure to understand what was called a nebulous concept by some, the answer from a former principal was forthright. 'They knew, *they knew*. But they did not want to go along with Berry. They would shilly-shally, dilly-dally. They did not want it to work'. Mr Berry stated that he had always been aware of the covert negativity of some and had simply chalked it up to the resistance and to the difficulties of change. However, he said that as the process worked, these persons lost some of their options. Others were confronted, not harshly, but directly. Three other persons reported that 'Berry always knew what was going on'.

Those who remained in the cadre could criticize openly. There was no penalty for doing so. Part of the process was developing an openness in discussion where disagreement was possible. An administrator said that he always tried to dissent as the loyal opposition, internally and in private, confident in Berry's graceful acceptance.

The process we have just described moved out from the superintendent and board and involved the identification of a group of people who could work together with a certain openness and mutual trust. They became bonded by activities, interaction, sentiment and finally interdependence. The formation of that group was one major way in which the central office was changed.

Along with the cleaning out of some and the identification and training of others in the central office itself, the process involved an active search for potential members of the group who could be moved into the central office or into principalships as vacancies occurred. Berry spoke of keeping a list in his head of administrators which was divided into three parts — those succeeding, those improving, and those who, if they did not improve with intensive help, would need to be moved out of their positions.

The search for, and identification of, new members for the ranks of administrators was not a one-man or a one-way process. The board president told of being actively involved. Berry saw the role of leadership in the district as centered in finding people who could work in the new flexible way. He and the president both thought that an organization could be structured in a number of ways if the people in it were able to assume responsibility and to share the decision-making. Berry said, 'I saw my role early in Riverside as finding ways to identify people who I viewed as being able to do that, to function in that way'.

To that end the district operated a sophisticated training program which included pre-screening and a semester or more of work. Activities included social affairs where Miller, the board, Berry and others were able to be deeply and personally involved with those aspiring to administration. A principal reported that 'Berry really did his homework. He knew his administrators'. Berry stated:

> We got to know them quite well and they got to know us quite well so that there was a mutual choice rather than just a one-way choice. So when opportunities came and we were to pick people for assignments, I had strong feelings about affirmative action in terms of opportunities for women and opportunities for minorities so whenever these vacancies occurred, I would lean in the direction of recommending appointment of people who represented those minorities.

Part of the process of decentralization was an active search for minorities and women to fill line positions. The data show that the district included a large number of minorities and women in positions of leadership earlier than in other districts. One minority person reported:

> Ray, in terms of minorities, began to get a lot of them in powerful positions. Of course, those were the years when there was a lot of turbulence but *he faced it* and he, I'm sure, had a lot of complaints from principals, but I think he was committed to equal education.

Berry's commitment to equal education was part of his larger intent to decentralize out of which arose his conviction to desegregate the district. He stated that, if you identify a minority person who will make a strong leader, 'don't play games. Hire that person and place him in your strongest majority school'.

Though parents did not know about the cadre of managers surrounding the superintendent, they were aware that leadership positions in the district were carefully thought out. One parent said:

> I think he chose his principals carefully. I don't think it was a political move with him and I don't think principals were chosen because they were from the right families or because the women were pretty. I don't think he did it that way. I think he chose according to ability and warmth.

The search extended nationwide. The board president said, 'I know we did a lot more searching and looking around the country for people. We did more of that'. Being brought in was only half the battle, because the evaluation process was as active as the search. Some who were brought it left and others were asked to leave if they were unable to work by consensus.

By changing the central office Berry hoped that down the road the working style would become a mutual exchange founded on mutual respect and mutual trust rather than so much of this 'I'm the boss kind of thing and you do what I say!' Berry reported, 'It worked'. Principals said those of the new direction in the central office suffered a real loss of power as they were forced to be subservient to principals and teachers in the clusters which were formed later.

This was the turn-around for which Berry had worked. He said he wanted the central office to become involved in servicing rather than commanding. He knew there eventually had to be a peerhood of all persons in the district but that could not take place until the central office had been flattened. A former administrator said it experienced a metamorphosis in its thinking, behavior, and relationship to the rest of the district. While this was happening, Berry began to challenge the principals.

Step 5: Challenging the Principals

The next step which moved decentralization to principals was as well planned as the formation of the central office cadre. The process needed more than superintendent, board and central office involvement. The key involvement of 35 principals required changes in the image and role of the principal. Self perceptions as well as those of the central office and teachers required examination. Principals were subsequently challenged to present and act on their own ideas concerning instruction and curriculum.

The first challenge to principals took place before Berry became superintendent. He used what he called the Hawthorne approach of giving money to principals who would work with their faculty to develop projects to improve instruction in their schools. The projects were funded on a per pupil basis and funding varied as much as $20.00 per pupil. Prior to those projects principals were dictated to as to what they could and could not do with programs in their schools. The program of projects deliberately set out to change the attitude of principals about what they could and could not do. Projects had a powerful effect on the attitudes of principals and on the attitudes of others about principals in the district. The intervention was successful in freeing principals. They began to make decisions at site level. Thirty or forty satellite programs resulted from this opening up of principal and staff creativity. The projects were eventually drawn back into the mainstream but not before they had contributed to a great deal of excitement and zest and to a vast change in principal status. Berry called it a powerful learning experience for all involved. The central office viewed principals and teachers differently, too, since they had witnessed decision-making and implementation at the site level.

The second challenge helped principals develop new norms. It could be termed a resocialization process. A principal said, 'Berry apparently recognized some of the skills and qualities in the administrative personnel who had just started in the administrative program in Riverside which might help direct the process of decentralization in the district'. As a result he had six to ten prospective principals work directly with him as associate superintendents on a number of projects in the district. Some spent their afternoons working in the district office. The close working relationships which were developed on these projects were a way to evaluate and instruct the aspiring hopefuls further on the concepts of decentralization. Here again we see instruction and a close working relationship coupled with example as instrumental in the spread of the process. The process involved shared feelings as well as shared decision-making (see Peters and Waterman, 1982).

The third extension of decentralization was made by engaging and challenging a very small group of elementary principals who showed certain characteristics. One man, part of the group, described the group as fairly young and active and aggressive. He said that Berry took the lead and that, 'He called in approximately half a dozen of us that had just been assigned to principalships in the district and indicated the direction he desired to take'. The data show at this time that probably while the central office had the concept of decentralization intellectually, they hadn't had much of an opportunity to practice the concept extensively. The group of six principals called together by Berry was to begin giving the central office a chance to practice decentralization of decision-making, though perhaps neither the organizational leader nor the principals fully realized the turn events would take.

One must be aware here that what seemed to be happening was that while the central office was being cajoled into and taught about consensus decision-making, a lower group of principals which ordinarily in all school bureaucracies, and heretofore in Riverside, took its orders from the central office was being instructed, or at least encouraged, to initiate decision-making themselves. This would of course not sit well with any central office, even one which had had some instruction and encouragement to share the decision-making.

The principals had had a bit of practice in decision-making in their work on projects. But those were not a challenge to the central office directly because the projects fell into the category of special. The superintendent who began the action as instructor and advisor to both sides ended as arbitrator and referee. The object of the instruction to the principals was to equalize more nearly the power of decision-making, not to start conflict and not for either side to win at the expense of the other. The encouragement to the principals to make decisions would probably be called mutiny if it were on the bounty, and treason if it were in the army. The word in school organizations is insubordination and it means that someone down there is stepping on a toe of someone up there. The encouragement took form in the budgeting process.

The germ idea for the development of a decentralized budgeting concept came from the associate business manager in the central office before it was suggested to the principals. While Berry was still associate superintendent, the young associate business manager had gone to him with the idea and had asked Berry:

> Why can't we set up a system where we provide an allocation to each school? A pile of money if you will, equal roughly to the total they're already getting in all the individual schools and say, 'this is your amount of money; now tell us how much you want to budget for supplies and magazines and textbooks and other kinds of nonpersonnel items'.

He reported that Berry thought that idea would fit wonderfully with his own ideas of decentralization. The business manager stated that 'as simple as that concept appears today, it was a little bit different then'. What the manager had in mind and had suggested was not the more fullblown decentralized budget which later resulted, though the budgeting process was never without guidelines.

One principal describes how the budget got decentralized in Riverside, 'The first direction Mr Berry gave us was that in order to decentralize one needed to control the dollars. If you controlled the dollars, then you had a good opportunity to be able to direct programs at levels which insure control. Dollars mean control'. Every principal would develop a decentralized budgeting process. Instead of being told how much money they were able to spend, they were to call in parents and staff and develop a budget, within a formula, for their program.

Principals realized their budget autonomy through an incident described as the 'Captain Queeg thing', referred to by one administrator as 'the albatross they have hung around my neck all these years'. A participant relates it. These principals felt they had been challenged that in order for them to run excellent programs, they needed to be in control of a certain portion of their budgets. They reacted by challenging the central office. An elementary principal reported:

> This budget decentralization was a very traumatic step. We challenged certain areas that had never been challenged before on a variety of things. They were the business office, the instructional division, and the pupil personnel division. We demanded to know why we would not purchase on an equity basis certain pieces of equipment (electric typewriters) that other levels of the schools had. We made, probably, some very unreasonable demands and we caused what we called then a great deal of disequilibrium. To cause disequilibrium we basically raised hell in the district with certain specific administrators and assistant superintendents.

The principals realized that their actions had placed the superintendent in a precarious position. They were, too, but one said:

> I think he probably was in the most precarious position of all. Outwardly he supported us and made no bones about the direction that he felt that we ought to be going. Yet, he needed to support good, solid people who were threatened by us. We were threatening those people with their very existence — their knowledge and the traditions that they had grown up with — that they had learned from other institutions

and people. They *had* to be boss and they really thought that they were being challenged in that position.

They were. It was at this point that the principals turned to the superintendent. They felt they were answering his challenge to begin to make decisions. Now they needed help. They asked in essence, 'Now who are you going to support, us or them?' Here one sees the dilemma in decentralization. According to a principal, 'Berry stood very firm and he said, "Well, I'm going to let the chips fall where they may and let's go from there. But I do think we will be a better district if we disseminate some of the decision-making processes that we have".'

The rule which was developed from this confrontation in the process of decentralizing decision-making was that there should be no hard and fast rule but that each situation needed to be reviewed and judged on its own. The issue here, of course, was not electric typewriters or even the budget. An elementary principal said:

> The issue was whether we at the principal level were able to decentralize a decision that was *ours* to make rather than having that decision made for us at a centralized level. And that was the whole process. The electric typewriters were just the catalyst ... And it became a tremendous battle cry ... the electric typewriter. It was kind of ridiculous but it was the thing that finally broke the back of the budgeting process.

Next, the principals were challenged in another interesting way by Berry saying to them, 'How would you open up this district? All of you want to make some changes. Let's assume these changes are good for kids. What research do you know that says so? What background? What studies have you looked at? What proof do you have?' When the principals answered with what they wanted to do or what they had researched, he would challenged them with, 'Well, what holds you back? Maybe, you ought to try it. I'll support it and I'll keep the board informed'.

At the same time he made it clear that while he supported the principals in directions they wanted to go and the directions they thought good for students, that he would not babysit them. He said:

> I'm not going to go over there and stand between you and the assistant superintendent of business or of instruction. I'm going to support them, too. But if you have good logic in what you're doing, eventually maybe, we can get these things done. I'm not going to tear up my administrative staff. I want

you to know that, but I will lend you support — strong support in these areas. And I'll keep the board informed.

The principals were invited to do battle on the strength of their own reason and research. He supported and encouraged their right to demand and to present their case on behalf of the student. One must recognize that during this process the leader identified with the good intentions of each group and trusted that those good intentions would prevail for kids. One parent spoke for numerous informants when she said, 'What he was all about was kids'.

Ordinarily, one would expect chaos from a leader who tells principals, 'If you think your decision about an electric typewriter is good for the program, good for what you are doing, then go ahead and demand to buy!' And who at the same time made it clear in the central office that 'where you draw a hard line, you have to make 13 exceptions, so maybe we shouldn't have hard lines'. This leader, however, appears to have had a larger view than of the situation at the moment. He was thought of as gentle and as a person who could wait and ride out a storm. However, even a leader with a vision has been known to perish by fire! In this case the leadership was successful through example, instruction and challenge.

The process caused some confusion, conflict, and confrontation. While not all principals rose to the challenge of a new role, neither did all central office personnel rise to the role of service. But another increment had been added to the process of decentralization. The power to make decisions had been moved down another notch by the intervention of challenging principals, which at the same time curbed the central office and brought it closer to a service orientation.

Step 6: Cluster Formation

Berry intended to open up the process of decision-making in the district to teachers but he needed a way to do that. Berry said, 'Teachers needed to be able to contribute to decision-making in depth, into the district policies and direction and planning so that they would develop the motivation and commitment that I was after from the start'.

He said that he 'began to reach for ways to bring about this participation process'. As stated previously, he had tried leagues in a part of the district in the sixties with UCLA and Goodlad, but they had not jelled. In those leagues, teachers and principals from one high school and its feeder schools met to discuss concerns such as articulation. In retrospect, Berry decided that involving just a part of the district contributed to the league's failure. This time it would have to be the entire district.

Clusters, the revived leagues, began in October 1976. Principals from five elementary, one middle, and one high school, plus one teacher for each 500 children in those schools met together for four hours once a month with Berry. The district was divided into clusters around the five middle schools and Berry gave five mornings each month to the groups.

Officially the cluster addressed the problems which plague all schools. The superintendent becomes a figurehead rather than an educational leader. The policies become rigid bureaucratic law rather than guidelines. Motivated employees become frustrated automatons. The right hand never knows what the left hand is doing. Standards and status quo become goals. Change becomes a problem to avoid. Improvement bogs down. Added expenditures do not increase returns. Finally, the owners of schools become onlookers, and maybe opponents, rather than participants and supporters (Berry, 1978, p. 2).

All of the above were legitimate problems on which the clusters worked. But Berry had, in addition, one overriding goal for the clusters, to move the power to make decisions down and out to the teacher level. That would continue the process of including all persons in all levels in decision-making. The cluster was to offer teachers a means to become peers with the rest of the educational community. To decentralize the district to this point the teachers had to become confident, eager professionals making decisions at the classroom level. Berry wished to provide opportunities for them to develop a concept of the district as a whole, what the district was all about, and where the district needed to go. A number of strategies were necessary to insure that it happened regardless of resistance.

Berry institutionalized regular meetings with separate groups. Each meeting included equal participation of all members. Subsequently members from each group became members of clusters and the equal participation continued.

The beginning was gradual, for example, he talked about the cluster concept for several years. Then he had all the principals to his home for dinner one evening, and a principal reported Berry, 'talked about this and it is something that he had been kind of throwing out right along for the last couple of years, because it didn't strike any of them as new. But they were nervous about an intervention'.

Second, he ordered everybody in. He said, 'this time we did the whole district at one shot, and everybody got into the act and they were ordered in'. He had learned, he said, 'that when only one part of the district met the other parts of the district could chip away at it and not buy into it'.

Third, he did not allow district officers to attend the meetings at first. This was deliberately done so that the 20 teachers and eight principals would not be overdirected by central office people. It was a shock to central office people that they could not participate in the clusters unless invited. An elementary principal tells the story:

> At first the downtown people were not invited to the cluster meetings and they felt it was a very hush-hush, closed — both doors in and out of the room were always closed. They felt left out. And as soon as Ray picked up on that, then immediately the doors were opened because he wasn't closing them out.

This last statement demonstrates how the strategy worked. Mr Berry knew the central office felt shut out. The exclusion was deliberate, he said, 'to reduce this business of central office people overdirecting site level people'. When they were allowed to attend later, by invitation only, they were asked to sit outside the circle and to observe unless they were invited to speak. The lack of voice and the outside placement effectively de-ranked them during the four hours.

The fourth strategy was that in the clusters teachers who had been selected by their peers would have direct access to him and through him to the board. Their information would not come second-hand from and through the principals. This was Berry's way of equalizing power. Direct information equalizes power. Berry said that he felt that, 'By having teachers sitting in those meetings along with the superintendent, along with the principals, reduced the opportunity for principals to overdirect, or overcontrol or overinterpret or misinterpret'. Clusters changed communication. When principals and the teachers went back to talk to the faculty, they were going to say, rather than interpret what went on at the meeting. 'And that's exactly what happened', Berry said.

Through this fourth strategy of intervention in the vertical structure, direct communication was firmly established from teacher to superintendent to board. Through this innovation 100 teacher representatives, in groups of about 20, had direct access to the superintendent while in the presence of their principal for four hours each month. These teachers distributed minutes of the meetings district-wide so all staff was included vicariously. One principal described teachers stopping everything when a teacher reported, 'Well, Mr Berry said thus and so'. *All* teachers wanted to know directly what Mr Berry said.

The last plan Berry had in mind was that he would listen, participate, and convince the teachers that they were heard. He did this by seeing to it that their ideas got built into programs. One example was the new report-card teachers wanted. Teachers were saying their report-cards were ridiculous. A teacher describes what happened, 'So by golly, he got some teachers together and got representatives from all the schools. And after two years of teacher toil the teachers had what they wanted, the best cards they could produce, and Berry took them to the board'. A second way the teachers knew they were heard was that once in a while he did something spectacular for them.

For example, one time teachers complained that they could not work without walls in one building. They asked, 'What do we do about this?' And Berry answered:

> We'll build the walls back. The teacher asked, 'Well, how long will that take?' I said, 'I'll start tomorrow'. In my role with the kind of structure we had that was possible. Now you can't do that with every decision. Some decisions require time, like the report card, but to build a wall all you had to do was to tell the business manager to get his crew and go out there and put the wall back. And when teachers could sit in a meeting and say that and the next day the crew showed up, that's dramatic! That really caused them to say to their peers and to others, 'Hey, wait a minute. This works. Let's not play games with it'.

Berry's strategies to involve the teachers had results. Teachers were elected or appointed by their schools, had a substitute for their rooms, and took the responsibility of representing their school seriously. When they left the school, one principal reported:

> They had such a sense of professionalism. 'I have been given something to do for the whole school'. That is very important because until you get people to own a part of this tremendous responsibility, you are not going to get it to work. And it worked. And it was exciting.

The cluster meetings had a well-planned agenda but humor had its place. A principal reported that the superintendent had a few quiet notes to plug into the discussion such as, where negotiations were or what phone calls he had received or,

> some funny little thing that had happened in his office such as, would you believe that a parent had called the night before graduation to say that they had all these relatives coming — their kid just had to graduate — and could he just walk across the stage though he hadn't quite made the grade?

From a short opening the meeting would be thrown back to the teachers and principals sitting in a circle around the table to ask the superintendent questions. A principal remembered, 'They could say, "Mr Berry, what about this? Can you tell us a little more about it?" And he would *level*. I mean just *totally direct communication* with that teacher sitting there'.

Teachers spoke up. Speaking up is one way of participating. Arnstein's (1971) typology of participation considers being allowed to speak up as middle level participation just a little above tokenism. Actual peer level participation requires that the speaking up be heard and implemented. Teachers were heard. It took a while for teachers to believe it. Berry said that once it got going teachers began to contribute at a highly professional level with great vigor. Again, he said that it was responsible people meeting together on issues, talking them out and then further communicating together, mutually with the people they represented. That is what developed early with the teachers. The principals were another story.

The clusters were something new. 'Principals', one principal said, *'hate new things'*. They were threatened by the intervention and some of them reacted by showing fear. Berry said that:

> It was fascinating because some of the principals were threatened by this. They felt it was an intrusion into their authority and their responsibility — that they had to bring teachers to these meetings — that it really reduced their role and identity and they didn't like it.

There was understanding of their fears and jealousies at first. Berry told them he was willing to meet with them for a period of time to help them develop an understanding and to overcome some of their problems. But they were told to 'keep in mind that down the road somewhere you're coming all the way in'. The confrontation was not harsh but limits were set. The group through instruction and example was expected to grow in understanding. Meetings were held for them with the superintendent for about six months. There the principals could, without teachers present, express their fears and talk them out. Berry said:

> You see, these principals liked this relationship with me and this direct tie. I had emphasized that the superintendent is going to be directly involved with the principals on an on-going basis and that gave them status and they felt good about it.

Because of the clusters principals were going through exactly what the people in the central office had experienced. They were about to give up power. Principals viewed the cluster as an intrusion and an intervention into their relationship with the superintendent. Berry said, 'and it was! It was deliberately that'. The principals saw the incorporation of teachers in the clusters as peer placement. That explains part of their fear and anxiety about displacement. They recognized it exactly for what it was, an attempt to further reduce the vertical organization of the district. It would change communication patterns and they knew it — thus, principal resistance.

First, they delayed the start-up. After they were reasoned with for several months, the 'down the road time' came and the leader said, 'Here's what we are going to do. We are going to do it this way, and now. The choice is not whether you do it but *how* you do it'. True to decentralization in the literature (Fantini and Gittell, 1973), each step of decentralization in Riverside involved confrontation. However, in Riverside confrontation came only after a long period of communication and instruction; it was finally used at each stage with at least some recalcitrant professionals.

The second step in the resistance of the threatened principals was to beat the system. At first only one cluster of principals began to meet by themselves before the actual cluster meeting, and they would hash everything over. One principal said:

> *That* was not the intent. They were simply to set the agenda. But the principals were just a little nervous about sitting with those teachers talking big administrative stuff. You know, what is big administrative stuff? It's the nitty gritty of every day is what it is, but the principals were making it, 'Sorry, I've got to leave my office to go do administrative stuff'. It had gotten all out of proportion and this sort of humbled them all but they were nervous about it so they would meet to hash this out.

The result of their pre-cluster meeting was that principals knew each others feelings about all the issues and how they should be decided. This knowledge dampened the liveliness of the cluster meeting. Their tactics did not work largely because the teachers spoke up at the meetings. A teacher said:

> Actually they were all surprised at how effective it was. They really didn't think the teachers would talk up like they did. And *none* of the principals ... the first year ever really spoke up and said what I knew they were thinking. They were just kind of like bumps on a log.

High school principals practiced covert participation. At first they ignored the clusters. One high school principal said that they had not attended for about three years because they did not feel that clusters would work, and 'we did not want them to work because they were a threat to us'. Then he said they noticed that the clusters were making decisions that had important input to the board members. 'Then, we felt *we had better go*'. The power the cluster groups developed had forced the high school principals in.

Berry gave people authority to make decisions in a decentralized manner. When that was understood by the district most subscribed to it. The clusters made decisions. 'There were forty things going on here at once', one central office administrator reported. And a principal said, 'That is risky for the leader! When an organization has forty things going on at once, there is risk involved for the leader who has given away responsibility in the form of letting other people make decisions'. Berry said, 'I had started the process and I had to go with it'. That is the element of authenticity which must characterize a successful leader according to Lee Roy Brittenham (1980). Once participants know they can make decisions, they will make them. And they are not all good. 'But the greater part of them must be honored, unless they are too off the wall', one administrator stated. Berry said, 'Once you start this kind of thing you have to be prepared to go with it, lead it, and make it work. You can't go to those meetings and make speeches and walk away. You have to go there and listen and participate'. One principal said that, 'in order to allow the tremendous growth of creativity in the schools that Berry was after, he really felt that it was necessary to allow a lot of things to go on that normally wouldn't be acceptable; and it was scary, and it was risky, and it was daring on Berry's part because he was in a precarious position'. The principal continued:

> This was the risk. It appeared that there was a sort of *laissez-faire* attitude. Every school could just go do what you wanted to do. 'You've got the hammer'. I think this was a deliberate move on Mr Berry's part, and it was a risky move on his part, very risky, that we be allowed the freedom to function this way.

One administrator said that there was the risk to the leader that it could go out of control and that it was hard to keep in balance and,

> If you've set it in motion, you have to abide by your rules that people can do differently. In time you must pull back together prior to reaching a chaotic condition — or you have confusion that goes in all directions. And in any organization with limited resources there comes a time when you must conserve the energy.

Berry became aware that the process of decentralization is circular and that it must not get to the point of chaos.

Step 7: Centralization

As decentralization proceeds, it is possible, as one administrator said, 'to get to the point where everyone thinks he is a baron and runs a baroney. Everyone is running his own private fiefdom'. Riverside did not get to that point but there comes a time when you must take the 3 out of the 17 programs that have been developed and get on with the task. Riverside curriculum experimentation was reined in at least twice in the period under study. Berry recognized that you couldn't have 30 or 40 satellites going in that many directions in a school system, but he thought,

> Still it is a way to get the door open. And I thought that down the road sometime we would pull this together and combine the better elements of this into a district plan and program of scope and sequence. And that's what we ultimately did.

The Riverside Instructional Support System (RISS) was implemented as one of the better curriculum experiments, and it was regularized and used throughout the entire district. *Man, a Course of Study*, was a social studies curriculum which was implemented district wide, and the district completed a science K-8 scope and sequence in 1978.

Informants reported and understood that decentralization has a process of intensity and relaxation. There was always leadership which took the ultimate responsibility for guiding the organization. It never became consensus decision-making to the point of anarchy as is documented by Sproull *et al.* (1978) where leadership was so democratic that it expressed no opinion. In that case all action stopped or never started because those skilled in competitive process, such as talking, fail to make decisions or to find a direction. The only action Sproull records was the continual setting and resetting of goals to the point of exhaustion for the organization. In Riverside there was a framework and a leader at all times, but there was freedom and autonomy within those parameters. That is how decentralization worked.

One administrator said it was Berry's design to free the district to begin with and always to leave it based on freedom; drawing it together after 18 years served to maintain balance.

A difference of opinion exists about whether the district was ready to move to more structure in 1978 when Berry retired. It was said that regardless of who had followed the decentralizer — or even if he had remained for a longer period of time — it was time for more structure but not centralization. One principal said, 'I think everybody had had a great deal of freedom for a period of time and they were ready to move to a more structured set-up. Not rigidly structured, but more structured'. However, another elementary principal stated that:

> It seems to me a lot of *zest* has gone out of living. Not necessarily because people want it but because they've made some choices that are leaning to a more centralized form ... They are not so much dreaming dreams of what they want any more. They are just hanging on to what they've got. And that always means that you let the other guy do it. And there doesn't seem to be much time for dreams or for implementing dreams.

A certain kind of nostalgia for the exciting participation under decentralization before the new superintendent came in 1978 was almost palpable in the district.

All persons agreed there had been recentralization since 1978. Opinions varied as to how much it was needed or how far it had gone. While one district officer mentioned that 'I haven't thought much about that concept, decentralization, for a long time', which would indicate that it had ceased to have center stage, another simply said, 'the commitment to decentralization was not there after Berry left'. While some stressed that at least the curriculum needed to be more centralized, most felt that the new superintendent was a real centralizer who did not favor the decentralization in the district. A principal reported that the superintendent who followed Berry 'was not in favor of clusters but in terms of decision-making in the district, there was no way for him to do away with them'. Another principal said that 'when the new superintendent arrived, there was covert conflict sort of like two stray dogs eyeing each other and wondering specifically what was going to happen'. What happened was that the process which had been set into motion was so strong that it held to a large degree. It was pulled up a bit, but not destroyed. A principal reported that the new leader had enough belief in the decision-making process and enough intelligence and 'savvy' as a leader to recognize that there were successful schools functioning decentralized and to leave them alone. This principal also knew enough to say to everyone:

> These are the directions that we are going and you need to select the best out of those and take them. But we are going in those directions. That means that we are going to be in a basic text, all of us. And we are going to have a similar curriculum. We are going to develop a course of study on a decentralized basis by committee structure by having input. There were enough people still that felt strongly about our decision-making process that we resisted any kind of centralized approach to the kinds of decisions we were making. So he went along with that. He said, 'Okay, I can work with that'.

Others reported, however, that when a centralizer became superintendent, there were those who, though the effects will never leave, reverted very quickly. An elementary principal stated that:

> When the leadership changes and there isn't the same kind of person, and strengths are in different areas, then ... they had to put it down on pieces of paper and have it in graph form, charts, so that we all knew precisely where we were. Then, very easily the principals sat back and said, 'Send out a letter from downtown for all of us to use. Or, let downtown do it and then we'll just be here to implement it'.

One district officer felt that even though the district became more centralized, the effects of decentralization carried over into the new administration because they had been widespread almost at a grass roots level and all had not been lost by any means. Another district officer thinks Riverside is now entering Stage III, as he calls it, of decentralization:

We have lost sight of the fact, and I think it is something that needs to be remedied very soon, we have lost sight of the fact that our staff has probably turned over 25 per cent in that period of time. So we have individuals in the district who don't know beans from apple butter about the decentralization process in this district. So, I think we are now at the third step in the whole process. The first step was for him getting people to understand what the process was; the second was establishing mutual trust and respect and trying the process. That is, we need to return to review on a districtwide and perhaps a community-wide basis, what we are doing now about decentralization. And do we need to re-educate some people who weren't in on it? Do we need to make some changes that now would seem to work better a different way after this experience? Not to go through and shake the whole thing up, but just begin to look at it part by part. And I think that's the step that we're in now and when that's complete then I think it will be good for another ten years.

Since decentralization is a continuous process, it does not work in the sense of having direct outcomes. When one asks, 'Is it working?' one means, 'Is the organization staying decentralized?' Because the dynamics of an organization are toward centralization, the success of decentralization is simply to keep it in balance and not allow it to become centralized. Interview and document data seem to demonstrate that the decentralization process has a life and power of its own once successfully launched and used in a school district; that it can withstand the loss of its visionary creator; can withstand a centralizer; and has proved itself attractive enough to warrant continuation through a third change of superintendents. However, since this study covered only the period up to 1978, the period of decentralization, it is not at present known how much recentralization has taken place.

Intentional Decentralization

Decentralization in Riverside was intentional, inclusive, and incremental. As an entity in one man's head decentralization was intentional. The man had a dream. Berry said:

I don't know why. I can't really say why I felt this so deeply
... Somehow I was almost instinctively driven in this direction rather than having to respond to somebody's direction or

pressures. The pressure were there. I recognized them and I still do but when I first started it was though I just — that was what I believed — that's what I conceived — and I just set out to do some of these things because it was natural.

Brittenham (1980) says the leader must start with a vision that will give 'structures to the progress and help to resolve potential problems associated with the envisioned change' (p. 153).

In order for Berry to actualize his vision, he had a project. The project entailed being in the system and being of the system but his intent was to act upon the system to change it. He intentionally 'tinkered' with that system to make it match the dream. He tells us what he dreamed of accomplishing with that 'tinkering':

> I really believe and I still believe very strongly that the roots of
> our country, and our whole concept of democracy is based
> on the strengths and power of individuals. And, in order to
> release that power you have to place trust and faith and con-
> fidence in people and let them know it and give them room to
> exert their power and their capability and their creativity with-
> in some defined framework. Define the framework and pro-
> vide the resources and do the monitoring and make it possible
> for people to function in that manner.

Berry let people exert their power and capability and their creativity within an organizational framework in a school district. He changed a highly centralized bureaucratic system without destroying it. Consistent with what is known about the best in leaders, Berry realized that 'though organizations are not perfect instruments of human intentions they are, however, a means of discovering better intentions' (Sproull *et al.*, 1978, p. 10) because the issue is cultural integrity, of actualizing rights which are implicit in the social order. (See Winter, 1966.)

The intentional acts of the superintendent were to change the central office from a command to a service structure, flatten the organization, give the principals site budgetary and programmatic autonomy, and allow participation in organizational decision-making. Consistent with Burns' (1978) definitions of real change, 'as a transformation to a marked degree in the attitudes, norms, institutions, and behaviors that structure our daily lives', the Riverside Unified School district was transformed in significant ways. One man said it turned from 'managing a district to teaching children'. The attitudes became ones of mutual trust and respect. The norms reversed, teachers were making decisions. A major change was to move decision-making to all the levels of the organization. To that end, he made many internal changes, including the central office, the principals, teachers, students, and parents. The result was a decentralized district where the teachers through clusters, became participants in the instructional and curricular decision-making process.

Part of the changes which took place emphasized excellence. One principal said Mr Berry had come at the right time to set the tone for the district because this district wasn't just any district. It was going to be a district that is out there doing its very best. Everyone was to be encouraged to buy in and take part in this responsibility. Excellence required the dedication and commitment of all people in the district. The quest for excellence established a climate of cooperation.

Conclusion

Decentralization in the Riverside Unified School District was originated by an intentional leader. Fundamental changes in the central office occurred early in the process. The number of positions decreased. The type of positions changed from controlling to servicing. The titles of positions changed from 'director' to 'coordinator'. Instruction assumed primary importance. The number of positions in the instruction division of the central office outnumbered those in the business, personnel, or administrative division by at least three to one. Central office emphasis changed from management and commanding to instruction and service. Consensus decision-making was taught and practiced. Norms and values consistent with intentionalism were transmitted and assumed by central office personnel.

Site administrators were granted autonomy. Identification and training programs for potential principals and site administrators were developed. Members of the central office, the board and others were involved in the program. Principals were freed from central office meetings to become instructional leaders in their schools. Budget autonomy was granted to the sites. Principals were challenged to make decisions.

The intentional leader developed strategies to include teachers in the decision-making processes of the district. Communication patterns changed. There was an increase in communication. Communication became two-way. Communication was encouraged early in the policy process. The intentional leader sought status equalization by the institutionalization of the cluster system. Dissenters were given time to change their views. Eventually, every school was ordered into the new organization. Teachers outnumbered administrators in the group. Participation by central office personnel was by invitation. The superintendent attended, listened, and participated in the cluster meetings. Teachers and principals participated equally. Teachers responded with increased input and commitment. Teacher input was developed into instructional programs.

The major focus of the school district was in the instruction of each child. There was an emphasis on teacher education. School rooms were desegregated. Dollars were equalized among schools. Curriculum experimentation was encouraged. Rooms and schools were redesigned to accommodate teacher experimentation. Parents were involved in classrooms. A new report-card was developed. The board sought input from all groups before making or changing policy. Teachers and principals made the instructional and curriculum decisions for the site. The decision-making process included all levels but the greater part of the decisions having to do with the site took place at the site. Parents were invited to take part in the decision-making at site level by supplying input to programs and budget development. Students were asked to give input to the board.

This work, unlike most which report the outcomes of decentralization, has reported the process. It emphasizes three areas: the intentionality present in the process of decentralization in Riverside, and its implementation, and its effect on the participants. The report is historical in that it focused on the process at it unfolded. It is historical though few, if any, of the details or happenings can be given a precise date like July 4 1776 Processes are nebulous stuff and much of the process we document is nebulous only because it was slow and evolutionary.

Chapter 4

Organizational Changes Inherent in School Reform Efforts

Levy (1966) claims that centralization at its extreme is when a single individual is able to make and enforce all decisions as to what each and every other member of an organization, society, or unit would do at any and all times in any and all contexts. In contrast, decentralization at its extreme would require every individual to decide what to do and when to do it without reference to anyone else. Methods for coercing and influencing would not exist; however, neither could a society survive if decentralized to this extent. Consistent with Levy's definition, the decentralization of school systems would enable groups heretofore excluded to be *in power*. The second basic assumption was that if *power* changed hands *children* would *learn*. Finally, decentralization was to occur as a result of internal or external imposition.

Riverside was decentralized by the superintendent, as a result of internal intent. The literature on decentralization efforts (see Wissler and Ortiz, 1986) reveals that decentralization may occur in two major ways: organizational members changing the internal structure of the organization or external parties participating in organizational affairs. Most decentralization reports describe external party participation. This participation, however, did not include decision-making. Decision-making was retained in the organization by original decision-makers. Administrative decentralization as reported in the literature occurred in two major ways. One was to create regional or area administrative posts. Most studies report this type of effort (Gittell, 1967a and b; Rogers, 1968; Sizemore, 1981; Hughes and Bratton, 1978; Cawelti, 1974; Patterson and Hanson, 1975). The second way to decentralize was to grant principals school site budgetary autonomy. As was reported in Chapter 3, the superintendent of the Riverside Unified School District decentralized in this manner. Chapman and

Boyd (1986) also report efforts to decentralize in this fashion in Australia. No other studies seem to be reported.

The movement to decentralize was associated with school reform. The primary aim was to improve student achievement, particularly that of poor and minority children. The literature reports efforts to initiate decentralization and subsequently link these efforts to improved student achievement and performance. These linkages failed to materialize in most instances. In the case of Riverside, the superintendent began with the efforts to improve the education of all children, and this intent resulted in decentralization. As described in Chapter 3, after 18 years of making changes in the school district, the resulting structure was a decentralized organization. This decentralization process was dependent on the personal characteristics of the superintendent and the types of changes which resulted.

Decentralization

The most fundamental organizational change in the Riverside Unified School District was that it changed from a bureaucratic to a decentralized organization. Decision-making was extended to all levels. The types and degrees of decision-making were determined by the kind of participation: individual, as part of a work group, committee, and cluster. It will be necessary to examine each group in order to determine how it was affected.

The School Board

The board of education changed its decision-making process in three ways: activity level, philosophy, and openness to input. The president of the board describes how the rate of activity increased:

> Our board was much more involved in policy starting from when Ray Berry and I would begin to talk about something at lunch and we'd think we really ought to try to be doing this and let's either talk to some people about it and it would begin to fan out. Or he would come to me and say, 'All right, we've got to do this or' ... Not just myself but all the board was more involved. We tried to have policy emerge with more people being involved from all kinds of groups.

Each member of the board and the board as a whole became more active than in past years. The board president said that prior to this period the board 'was pretty much a rubber stamp and handled some functions', but they were not policy-makers. The change meant the board became policy changers and makers. Policy was no longer handed down. Instead, they constantly and actively made every effort to get all the input they could. He stated that they became particularly interested in input from poor people and felt strongly that there was a lot to be gained by discussion and involvement of other people, as their policy changes emerge. He explained, 'We sought to do everything we could to try to bring in and to recognize that not only do we need help in making decisions but also in our ability to carry out those decisions'.

In the centralized district people were not involved early. Under decentralization, people were involved early in the policy process. This early involvement was both personal and organizational resulting in both kinds of changes. The personal change described was one of greatly increased activity. The effects on the school were two-fold: policy was not being handed down, and the search for a new kind of school personnel was begun. A board member said they were looking for people who could assume responsibility and make decisions without a rule book. They needed persons who knew when to make a decision and when to call for help from others. As the board became more active, the harder issues of education were addressed. The responsibility was not only to know how many graduates went on to college versus how many did not but to address the tougher issues and problems that come with providing education for all children.

In summary, the board became more active. The system formulated its policy based on early, wide, public inclusion. The organization searched for a new type of personnel. Additionally, the harder issues of education were faced in order to provide education for all regardless if they were college-bound or not.

Central Office

The process by which the central office operated changed because power moved from the central office to the school sites. A principal said that the movement 'grossly limited the authority of the central office people which they felt they firmly had', thus affecting its communication pattern, its work style, its procedures, orientation and morale.

First, decisions were made with more discussion than had previously been the case. An administrator said the change in the central office was from 'passing out edicts' to trying to discuss things a lot more, and to think and to talk about problems that might arise before something was put into place. It was a move from passing out bulletins of 'this is how it will be' to a consensus style of decision-making. The result was that everyone was encouraged to participate. Although encouraged to participate before, administrators were challenged with, 'Here's a problem. What would you do with it? What do you think? And how would you implement that?' As the central office moved from vertical to lateral communication as a result of the process, mandates were replaced with cooperative influence. This represented a change from the days when principals *stood* in front of central office desks in response to summons. One district officer reported that the work became more enjoyable as these changes occurred.

Granting principals budgetary and programmatic responsibility greatly reduced the central office's influence. The central office's responsibilities became those associated with instruction and children. As one official stated, 'That operation addressed the education of youngsters and that is what Riverside is concerned about!' The central office had experienced a significant shift into an instructional support system. The central office focused in on business only as support to the instructional arm of the district.

Because of the focus, the central office became involved in aiding, rather than controlling, the daily work of the school. For example, one central office administrator said he became a facilitator and implementor for the teachers and the principals in the cluster system. Schools and personnel connected through the cluster were able to have instruction and curriculum effectively coordinated. Central office administrators listened to, served and planned *with* school site people.

When decision-making changed, central office personnel felt the importance of a proper work performance evaluation system. Prior to this period evaluation meant only a visit or two and the checking off of an evaluation form. The new concern became, 'How can I help this individual? I will give whatever help I can but at some point — after giving as much help as we can — we have to state that this individual is or is not going to be able to function in the organization appropriately'. District officers said that the effect of the fundamental change in decision-making in the central office was an enhanced atmosphere of trust and productivity.

In summary, the process as it involved the central office changed communication patterns inside and outside the office to one of listening; changed decision-making to consensus; changed work style to facilitation; changed orientation to a primary focus on instruction; fostered an acceptance of the importance of work performance evaluation and enchanced the atmosphere of trust and productivity. They were never 100 per cent in place and totally operative but there was an intention that they should be.

A central office person said about decentralization that 'It becomes a vehicle to solve the problems that you have. The decentralization process itself is based on some kind of trust in fellow workers and in *knowing* that the intent to succeed was there'. The superintendent stated that he discovered people began to feel comfortable with this new structure and their reward was the invigorating experience they received from being involved.

Principals

The results of principals being challenged, creating projects, decentralizing their budgets and battling for the right to make program decisions were that some principals held a new image of their role and that the central office viewed them more as peers. The new image and role were needed because while principals had once been told what to do, that was no longer the case. One high school principal told how it used to be. 'I used to clash with the central office before decentralization because the central office position was that it hadn't been done that way before, we never do it that way and *that is not* the way we do it'. The board president described the changes with principals:

> Decentralization gave people in decision-making positions the authority and responsibility to carry a program or a school and to make the decisions that were necessary. It was a movement away from a bureaucracy where policies are ground out at the top level and handed down to people to be followed and if followed no matter the result you're okay and you've covered your hind quarters.

Decentralization meant leadership to the principal in the areas of budget, curriculum, instruction and evaluation. Budget decisions were theirs within limits. What would they do with it and their new power? They had at last an opportunity to control programs through the budget.

For 'retired-on-the-job' types and even for some excellent principals it was a time of terror. They were faced with a set of demands which were 180 degrees different from those to which they had been socialized both in school and on the job. Some of them had been principals in the district for 15 years. Now they were faced with decisions such as: Will I be able to work in this way? Shall I stay? Will I risk losing control of the people under me? What if I stay and fail? The story in the district office was that sometimes decentralization stopped at the principal's desk, either because he hated the new direction or because he reveled in it and would not share. There were the old hands who resented it. For others having power to make decisions was heady. Principals who had been ground between the central office, teachers and parents and had often felt 'Why do you need me? I can't make any decision anyhow?' became so powerful with money for the first time that things got sticky with some of them. A district officer explained that:

> the principals would decide a function didn't need to be done so they would say, 'I'm going to fire my bookkeeper, and I'm going to take that money and use it to buy textbooks'. Well, that's great in theory except then the book work was not done and it was a disaster. So we inherited this incredible mess. Yet, the principal had all this money and was a hero with his staff.

The new process required a balance between power and responsibility. While the role of the central office was service, it was not entirely advisory; just as while the principals had budget decision-making discretion, it was not without limits. Some principals recognized and accepted the balance. Others did not as principals themselves reported. By November 1970, 17 out of 21 managers responded positively when they evaluated decentralization in the district. Most said there were opportunities for principals to inspire, create, perform, and serve. The consensus by that time was that decentralization was a help to principals. One principal said, 'Decentralization gave them new life. It was just new zip. It helped principals who used to turn and say, "You know I can hardly make a decision with all these rules and regulations coming down to me"'.

Berry's hope echoed Gross and Trask (1976) that principals should become instructional leaders. They were 'freed from 1000 meetings to be in their schools', one principal said. Another said it was their opportunity to sit down with their staff and parents and ask, 'What are we doing and what should we be doing?' Before decentralization schools had been as uniform as possible even down to the same recesses. A high school principal said after decentralization they were encouraged to recognize each school had ethnicity and an economic makeup bringing a unique set of problems and they were encouraged to look for a different set of solutions. With reasoned argument and research and with input from staff, principals could try new experiments in program and instruction to meet the unique needs of their communities. Their school did not have to be like any other school. This was an important point of pride to each principal interviewed. An elementary principal reported that:

> It makes you feel that your school can be just a little bit different and it's OK as long as you have your rationale. Then you do your very best because you're not trying to follow some model that somebody put up there and said, 'This is the way it is going to be'. You're saying, this is a good model and I'm going to take a little bit from each. But it's going to be this school because I know this school. So nobody's competing. My school doesn't have to be *just* like that school over there. My school can be just right.

A third principal said the results were unique programs that sharpened teachers and principals so that their views were not so directed, narrow and staid.

Emphasis on evaluating programs and personnel gained importance. Principals were freed to be in their schools and their classrooms in order to evaluate teachers wisely. An elementary principal said that, 'If you were going to be able to evaluate teachers in and out of the system, and there were many that needed to be evaluated out, the only way to do that is to be in the classroom'. Decentralization was to spread and to share responsibility. A district officer said that, 'It really pinpointed responsibility at all levels'.

In summary, the effects for principals of the changed decision-making process meant they were responsible for their schools. They were to make decisions. They were to decide and not pass decisions upstairs. With research they were to present their plans to their peers and eventually maybe, get all those things done. Their schools no longer had to be alike and neither did their programs. They were no longer to face only towards the central office and the school board but could turn towards the staff, children, and parents to serve a unique set of needs. They were given an opportunity to share the special programs in each of their schools with the board once a year. Board minutes demonstrate the variety of the programs and the warm reception with which they were received.

A further result of principal leadership and autonomy was that it was tremendous for morale. A loyalty to the district developed. One principal reported that he and numerous others had stayed because they never could find the freedom and autonomy that was theirs in Riverside. He reported that at one time he had thought of other things,

> but it made a lot of difference to find out what schools were doing in other districts. I found they were all looking to Riverside. And yet the other superintendents did not want to take the risk of decentralizing. So, I felt I never really wanted to leave this district. A number of us have not left because of this.

The data show people have remained in the district. Others who left are nostalgic about what they had experienced and have now lost because they have been unable to find it elsewhere. One principal still in Riverside remembers the time of greatest decentralization with wonder, 'It was exciting to know that somebody over there thinks you can do it', he said.

Teachers

The goal of decentralization was placing decisions as far as possible at the classroom level. The hope had been from the first to decentralize to the level where teachers had a 'say in what was happening to the curriculum, program, students, and to the parents with whom they functioned', one principal said. The fulfilled goal thrust teachers into a position of importance and responsibility. They were to make decisions.

Not all teachers were comfortable with the change. A parent reported that some teachers volunteered to leave demonstration schools because they felt like they were going to be looked at and would have to be accountable. Other teachers wondered why administrators were paid when they have to take time out of teaching to plan budgeting? Being expected to input into their school policy was a change some teachers did not relish even if it increased autonomy. But the teachers who welcomed the challenge had an opportunity to change work styles in the following ways.

First, teachers searched for ways to cooperate more with each other in teaching children. They communicated and worked together to help each other become better teachers. They shared techniques and teaching strategies through show and tell workshops. There was an increase in team teaching and math labs for grade 1–3 with three teachers teaming. Walls were knocked out, doors were cut between rooms and one school was built to foster teacher cooperative experimentation.

Secondly, teacher work style with children broadened as they related to and accepted differences in children. Some informants felt that students were involved in the decision-making process in their classrooms as the result of the teacher's enthusiasm for decentralized decision-making. A minority teacher commented that, 'teachers showed tremendous acceptance on their part', and she praised the sensitivity which developed to all children. This sensitivity destroyed stereotyping which had existed in the district before, regarding that some children can and some children cannot learn. With decentralization came the attitude that you took children right where they were and moved them as far as possible. One said, 'That was the attitude that was pervasive in the district as a result of decentralization'.

Teachers were stimulated to apply child-centered concepts by designing programs as they saw the unique needs of children. Teacher freedom fostered curriculum experimentation. There was an explosion of curricular creativity. A principal reported that in many instances these experiments were fairly routine, like rediscovering the wheel. But there was excitement when teachers, principals, and sometimes parents got involved in them. The process had power and it began to break teachers and principals away from the district structure and teach them it was *alright* to experiment.

Curricular experimentation involved everyone in some schools. Many methods of reading were tried and teaching materials changed. One experiment led to a method of testing students' mastery of basic skills and of providing rapid feedback to the teacher so that both profited. Physical education classes of therapy through games were developed. Physical motor skills were assessed and trampolines were used in some schools. 'These were extras on top of a full academic program in which wonderful things happened to curriculum for kids', an elementary teacher told us. 'While some methods and experiments were not the best and perhaps the major share failed, still the focus was on the student and unique things happened to kids', a district officer stated.

Teacher work style was reported to have changed with parents too. One teacher said that teachers could communicate the school's goal to parents when everything in the curriculum and projects was centered on the successful education of their children. The result was that teachers had success with the children, community and parents. Teachers received inservice training on skills relating to differences among parents and children. A district officer said the process changed teachers and staff by sensitizing them to a different mix of youngsters. Teachers, thus, took the initiative to tackle the problem of teaching all children regardless of ethnicity and social status. Through regularly scheduled meetings and increased interaction with parents, teachers were able to *draw* the parents in.

An important work style change for teachers was the way they related to administrators. The cluster organization provided an opportunity for teachers to sit in meetings with administrators as peers. The meetings served to inform teachers what was happening in the district and teachers had an opportunity to provide early, useful, and continued input into district policies. Administrators listened to teachers. A principal reported that teachers spoke up and their ideas were accepted by the superintendent who laughed but really meant it when he told them, 'I was hoping that I would be long gone before the day ever arose that I would have to do this again, design a new report card. But if you say so, alright, we'll do it'. Teacher ideas were not to be rejected without a trial run even in the face of questions in the mind of the superintendent. Berry said that he *admired* those teachers as they persevered with the project for two years even though they were chopped alive and crucified by their peers until they won the district over. The superintendent who had supported the teacher innovators against his better judgment then took their work to the board where it was adopted for the district. One teacher said that she saw mutual respect among teachers flowing out of the clusters. 'We learned to know that other people had problems such as we did', she said. She quoted a teacher who once walked with her into a cluster meeting, 'At last I have some power!' She explained, 'Now to me that wasn't why I was there. I was interested in facts and knowing what was going on not just in my little classroom but in my whole school and this was an opportunity for me'.

Through clusters teachers learned that they did not have to be alike. For example, they did not have to write on every single cumulative folder unless another teacher could benefit from what they had to say. A principal reported that the teachers reacted with amazement at first upon learning they could make that decision. Concrete decision-making extended to them served as a tribute to teacher judgment and an example of the trust on which the process was being built.

Teacher work style changed with the principal as they worked on specific problems and solutions. After decentralization some schools had all of their teachers involved in all decision-making for classrooms and curriculum. For example, one teacher claimed that before decentralization teachers in her school were told what class to teach; what grade level they would have; how many children would be in the class; and if they were going to have a combination class. This was known as the numbers game. After decentralization teachers were informed about their school enrollments and asked to arrange their classes to benefit all. One teacher reported that in her elementary school, 'The decisions were left completely up to the teachers. Sometimes a teacher would suggest what we thought was a good idea and that's the one we took for the whole school and everybody went along with it'. Teachers also had an 'opportunity to sit on the committees to select a new principal for the school and to help develop the policy on retention', another teacher reported.

Besides changes in teacher work style with each other, children, parents, and administrators, there were changes 'in terms of attitude, vigor, caring, commitment, and willingness to spend time', one administrator said. Another said that he saw a blossoming of leadership particularly among principals and teaching staff. He said this resulted from their experiencing the fact that they had opportunities for input and to effect change.

To elaborate further, a teacher described what the cluster did for teachers and the district:

> We would sit around and have a chance to talk and to listen to other teachers and other principals and interact with the superintendent. That just made me feel like *wow*, I'm part of a team here. And we would all get kind of going!

The only absolute for the teacher in this process was a decision to participate. Every teacher had the possibility of representation and input. A principal reported that not all teachers took part because it was such a unique idea and so unbelievable that a lot of teachers didn't practice it. At the teacher level there were persons who never caught the feeling of decentralization for whatever reason. Those who did reported personal effects and changed organizational work styles. A teacher in explaining how she perceived her decentralized district said, 'If somebody walked up to me from Sacramento or New York City and asked, "What is this district all about?" I would say, "This district is about personalized instruction and teaching every child regardless of race or socioeconomic status". We know it. *We all know it.* Teachers are confident in knowing what their job is'.

Another teacher said, 'When I came into the school and Ray

Berry was beginning decentralization, my whole being blossomed with him in developing this aspect of teaching children. I understood his goals and was supportive of his goals because they related to what I was all about'. Decentralization thus, provided leadership and an opportunity for teachers to experience what Maslow (1954) calls teacher's self actualization.

Parents and Community Members

Parents and citizens were included in decision-making in five ways. They were brought into schools to work on programs and budget. They were drawn into the communication cycle by parent advisory committees (PACs) which were begun in Riverside before they were mandated by the state. Every school had to have one and the leaders of those groups and principals met once a month with the superintendent and some members of the board. Berry said that in no way was the Parent Teacher Association (PTA) to be abandoned, the PAC was not a PTA type of organization. Like the cluster organization in the district, the PACs were groups of interested citizens meeting together on issues, talking them out and then further communicating mutually with the people they represented.

Parents were included through attendance at some cluster meetings when there was a particular issue that might interest them. They were also involved as aides and volunteers in numerous classrooms and as tutors in educational centers in the city. Finally, Riverside employed twelve community aides, as a resource to the community, during some of the time under study. They were to be in the community, riding buses with children, walking neighborhood streets, and available by phone to parents who had questions or needed information. Their purpose was to establish a connection between the home and school. Those persons employed were well known and well respected in their neighborhoods. Their skill and training added a dimension of service to the parents of the community which was original to Riverside. They also served as a witness to the intent of the district to include all.

Minority parents who had been included in various advisory committees prior to decentralization were also drawn into the schools by the integration of their children into majority classrooms. Mexican-American parents saw this inclusion in positive and in negative terms. Equating decentralization and desegregation, they believed that school administrators knew federal money was going to come to poor schools and that the administrators had deliberately deceived them by not giving them that fact. Therefore, Mexican-Americans felt they lost in two ways when they were included in majority schools by one-way bussing: they lost their own neighborhood schools, and the money they could have had to improve those schools. The parent commented, 'If our school had not been taken away, just think how much further our children could be today'. He also indicated that Mexican-Americans would have welcomed majority children in their school so that integration would have been equally borne. Thus, this parent was affected by loss of federal monies, next by the loss of the neighborhood school, and thirdly by the removal of his children who were bussed to numerous schools. He had no quarrel with decentralization, but he believed that teachers and his community were not told what they *had* to do and how the administration intended integration to work. Consequently, for him, the inclusion of Mexican-American children in Riverside schools had been partial and differential.

This parent also felt that decentralization and desegregation had taught him the reality of schooling. The reality to him was that there were different values outside his community, and they were values he and his group did not like and did not have, but ones they had had to learn. The Mexican-American parent said:

> So, when we see that we are living in that kind of society then we have to find a place for ourselves and always be constantly battling one another. And we have to learn as we go into these battles to be more effective.

While his group had once been excluded geographically, he felt that decentralization and desegregation likewise excluded them from the truth and from respect. Though not stated explicitly, this person had learned that inclusion in the majority society was a matter not of right but of fight and of conditions put on you.

The positive outcome of being even partially included in the process was that more Mexican-American children were staying in school and doing well. They were learning how to deal with the reality that could not be changed. This parent stated he was the only male in his age group who had completed high school in his community. He was pleased this was being remedied, but it was clear that in his opinion the price paid nearly equaled the gain.

A Black parent described changes which affected her family positively. Before decentralization she felt her son's school was closed to her, did not welcome her and whenever she went there, she felt that she was looked down on and made to feel like a 'dummy'. She was concerned about her first child and experienced difficulty getting information. She was made to feel by school personnel that they were the experts and so she had turned to the district office for help. She was welcomed and helped at the central office, but later berated by the principal who said, 'How dare you go over my head!' Following decentralization new principals were appointed, and in this parent's experience the school became open to parents. Principal and teacher attitudes were now receptive to parents.

She considered the parent advisory committees as the best thing that ever happened to the schools. Through them she was invited to help set up or give input on budgets; district personnel came to explain test scores; and classes were held which taught parents how to help their children succeed in school. This parent states that all of her children, including the first who had so many difficulties, have graduated from high school. One is attending the University of California, Riverside and another is enrolling for the coming term at the University of California, Berkeley. Decentralization is perceived by this parent positively because changes in the schools allowed and helped improve her and her family resulting in academic success for her children.

Knowledgeable about the instruction of the district, she described the LASER and RISS programs as instructional programs containing detailed profiles of children's progress. Individualized instruction for this parent meant that her children were receiving the best education available.

Parents were included in the decision-making process in five ways. Assessments of the effects varied. However, even those parents who cited negative results harbored no rancor for the decentralizers saying, 'Really, my feeling was that they were trying to change it to the best'.

The Establishment of Organizational Priorities

Child-centered concepts

The profound change in attitude that developed in Riverside was that everything else became secondary to that child in that schoolroom who was struggling to learn. Again, as in all other fundamental changes in the development and evolution of decentralization, not all persons understood. But many did speak of Riverside's child-centered concepts. One principal, two teachers and four administrators said in different words that the child-centered concepts were individualized instruction designed to teach each child to that child's maximum capacity. These are common goals for schools, however, most school districts fail to fulfill them for all children. In Riverside the focus attempted to realize our democratic rhetoric. Berry said, 'We were finally beginning to try to do what we have said in this country we'd always do which is to educate every child effectively. We never *had*. We hadn't really come very close'.

Berry believed that if the school's interest is on the child, there is *no* child who cannot be educated to the maximum. Rejecting the belief that some children cannot learn, Berry provided an opportunity for equal education. The onus was placed on the district. Riverside's leadership intended to accept the responsibility to educate not only the children who can learn easily, but all others. Through the process of decentralization Riverside assumed its mission to instruct children and tried to destroy the negativity of stereotyping.

The second fundamental child-centered concept was that we must cause learning to happen. Berry stated emphatically:

> Every child must perform in a very specific manner at that child's *utmost pace and capability*. So what we're doing here is to find a way to do that. Not to say each child progresses at his or her own pace. That's ridiculous. You've got to lead children. We're the adults. You've got to provide the direct instruction and all the other things that cause it to happen.

The district moved in at least five ways to cause it to happen. First, the district involved the community. At one time there were hundreds of volunteers in the classrooms. Second, it built libraries. In the 1960s there was only one library for elementary level children. Because of the new focus, libraries were established in each school. The intent was that libraries should be instructional and media material centers staffed by skilled resource persons in order to provide enriching programs for students and teachers.

Third, there was an intense search for ways to personalize or individualize instruction for children through cooperative teaching. One explained, 'Ultimately, it comes down to a teacher's knowing where each child is and doing something about it'.

An elementary principal said:

> And that was what Riverside was all about. Personalizing programs for kids. Now, we are *not* talking about one to one.

> We are *not* talking about every kid had his little program and the teacher spending time with that youngster specifically so many minutes per day. We *are* talking about regrouping, and testing kids and as soon as they become successful in one area allowing them to progress by encouraging them and pushing them. *Never* keeping them back for any particular reason such as convenience, because it's not easy to get to them.

New curriculum was written, developed and piloted in Riverside. The desire to know where each child was resulted in what one teacher called a very meticulous, almost too meticulous, way of keeping track of exactly every little skill that each child had in reading and math.

Fourth, the traditional classroom organization was affected. Riverside recognized its inadequacies. Berry said:

> You couldn't deal with the complexities of needs and the range of children's readiness and backgrounds and all the rest by thinking of a class as one year's worth of so much knowledge for 30 children. It was an outmoded concept.

Riverside moved to multi-grade levels, to team teaching, and out of those grew experiments such as math labs with three teachers and three aides in one room teaching. A third of the class would be in drill and practice, another third would be drilling on machines and the other third would be interacting with the teachers and aides. One teacher said, 'It was a rewarding experience to teach like that'. Dozens of experiments in teacher cooperation took place as teachers became educational decision-makers. And, as one teacher said, 'The scores went up'. Others confirmed that over the years performance was affected because teachers had a voice in their work.

There was a fifth change in the district. As was stated previously, before decentralization resources were distributed unequally. Decentralization changed that disparity. The business manager explained, 'Over the years we got them all equal. For many years not everybody was equal'. Decentralization proved to benefit the district by equalizing it.

In summary, the child-centered concepts in Riverside focused the district on realizing its primary mission of instruction and it moved in at least five particular ways to accomplish that mission. It was felt throughout the district that this occurred primarily because as one teacher said, 'The education of children was the superintendent's bottom line'.

Communication

Decentralization changed communication patterns in the district in five ways: amount, direction, timing, context and formality. The intense desire for input from all people displayed by the board of education under decentralization was also sought from the schools. Many informants stated, 'I think there was a lot more speaking up'. An administrator remembered constant communication throughout the district as well as in the district office during the period of decentralization. He said that Berry,

> began pointing out in many different ways, sometimes individually, in large groups or meetings, sometimes even in memoranda, certainly in board meetings that the human resources that were obviously present in the staff were not used effectively.

The participants also described an increase of two-way communication. Three devices in the district provided two-way communication. The first was by listening, the second was by an open door policy, and the third was through the cluster organization. Listening was an important part of the decentralization process. Listening became a district-wide trait which started with the board, for its president said that the board's natural instinct was to reach out and be open and unafraid to listen to people. The superintendent was also known throughout the district as a listener. A principal said, 'and *listen, he listened* ... he wasn't a rapper. But he would listen'. A parent said, 'He listened to what you had to say and if he could do anything about it, he would. We hadn't had that in a superintendent'.

Two-way communication was fostered by an open door policy. Parents said they felt free to call the superintendent. The door was open to teachers and principals too. A principal said, 'You didn't have to come in with a problem; his door was always open'.

Two-way communication was augmented by the innovative device of the cluster meeting which was recognized in the district as the most profound change in communication. Through clusters, direct two-way communication was firmly established between teacher and superintendent, principal and superintendent, and principal and teacher. Direct two-way access to the superintendent on a regular basis in the centralized system had been previously denied to teachers.

Two-way communication with its listening, open door and face-to-face contact contributed to reversing the direction of communication. It became bottom–up. One administrator characterized the change by saying, 'This was a reversal of the top down to a grassroots sort of thing ... and *edicts* were not handed out so frequently!' Communication changed from directives being passed down to asking for directions by the board and central office. A board member said, 'All the board members felt strongly that there was a lot to be gained by getting ideas and listening ... we were trying to get all the helpful input we could'. Parent input was developed by bringing parents in and sitting down and talking about program. And lastly, students were included as non-voting members on the board of education. Interview data show that their input was sought.

A specific result of the effective two-way bottom-up communication process was mentioned by a principal who believed that:

> One of the reasons why we went through the civil rights thing *almost* unscathed was because we had already developed a decision-making process that involved parents. I really believe it was because of the way we were set up to function with parents in the first place that allowed us to do that.

As stated previously, the district had a parent advisory committee in place in every school and these committees served to provide bottom-up input.

A crucial change in communication was in its timing. The board didn't just want a lot of input from everybody. They wanted it early. A board member said that the board couldn't let teachers work two years on something and say in two minutes, 'We don't like that idea'. The board wanted to input and to know what was going on early, and reciprocally they wanted to hear from the public and parents early. The board president stated, 'The community sensed that they could begin to be heard and listened to when whatever they could say would make a difference'.

Communication in the district became the 'creative conflict' which Follett's (1941) work describes as necessary for a productive organization. Everyone had a right to disagree and to say so. The disagreement did not have to remain private, for as one principal stated, 'It was good for everyone to see that the superintendent can really hit the table and say, "You know, I couldn't disagree with you more!" And once in a while people really got to that point'.

A board member said, 'We were not afraid to be told we were wrong and that doesn't mean that sometimes you didn't have to stand up and say no'. He explained that there were times when after listening and reflecting he had to say, 'I disagree with you and that's it. Somebody has to make the decision and we've made it'.

Creative conflict communication enhanced trust. One central office administrator said that decentralization established a process where dialogue was constant and impacted the entire district and community. He felt the availability of constant dialogue had affected student scores because teachers had a voice in what they were doing in the district. Likewise, he believed principals reacted positively as did parents because they were included in the interaction. He continued, 'So there is this open dialogue from the first employee to the oldest parent and everybody in between'. Trust developed because the input of people was honored.

The last way in which communication changed was that informality increased in both writing and interaction. For example, much of the process of decentralization was completed before anything was presented formally. As stated previously, Mr Berry persevered in preserving spontaneity and creativity during the process in spite of continual pressure from persons who were used to and comfortable following orders to get this in black and white. A district officer said:

> Writing it would have been the easiest thing to do but Berry thought that would defeat the purpose. He referred to it as giving orders to those not yet ready to participate in what we might call participative management. I think Mr Berry and several other administrators recognized that success required a genuine feeling of cooperation instead of one of compliance to imposed rules.

Communication about decentralization was intentionally kept nebulous so that people would feel free to add or to change as they participated. Communicating feelings, beliefs, or attitudes while transmitting information insured that the decisions arrived at were in line with instruction and the classroom. This type of communication required learning. 'Orders will not take the place of training' (Follett, 1941, p. 53).

In the usual hierarchically-structured organization, communication follows channels. In Riverside the kinds and uses of communication were different from those of an hierarchy. Riverside developed effective channels of communication which limited paper work. One officer said, 'We would have been buried in paper if we had written it all down'. Secondly, a vocabulary of shared words was developed over the years. Informants often laughed when they used one of the words like emerge and said, 'I bet you have heard that word before'. Riverside developed a 'virtual technology of keeping in touch' (Peters and Waterman, 1982, p. 123). This was established by the superintendent's regular visits to each school, by his open-door policy, by cluster meetings once a month, by parent advisory committee meetings, and by the search for input by the board. Communication was face-to-face with the superintendent, and as Peters and Waterman would say this was 'management by walking around' (1982, p. 67). Communication was changed to having teachers, janitors, cafeteria workers and parents sitting as peers developing budgets and programs. Communication was bottom-up, and when it did come from the top it was usually in the form of instruction instead of edict.

In summary, the evidence that communication changed in the district at the board, central office, and site levels represents a fundamental change in the district. Communication increased, became two-way and bottom-up, was welcomed earlier in the policy process, conducted in the context of mutual trust, often informal, and served to equalize social relationships by encouraging input from all.

Rules

For decentralization as well as desegregation there were few written rules which could be codified. People had to be able to produce under loose rules and regulation. The board president stated:

> You weren't going to find anything in a book about how to do this. There's no way that you could outline every situation and say, 'Okay, if this comes up turn to the book. This is what you do'. Well, then you've got to have a lot of good people in the school and people that can inspire the teachers with confidence, help them, and step in and make decisions when they were necessary.

There were rules but good judgement prevailed. There were even edicts on occasion. For instance, there had been an edict that each school would have a parent advisory committee.

Relaxed rules did not mean helter skelter or that management of the schools was *laissez-faire*. Instead, there was a balance sought to decrease the risk for the leader. A principal expressed this need for balance as the absolute of decentralization in the following way:

> Decentralization can probably take many different forms. I don't think it is a single thing. I think it is a broad idea. And I think it involves participatory management and it involves handing responsibility in a smooth and rather structured way

to people outside the central office. It is not being totally devoid of the central office because you don't want 35 entities out there. So the balance, the real crux of decentralization is keeping the thread, keeping the attachment and keeping things total. Getting the broad view. And, it really is giving everybody the broad view of the whole system so that they can go back and within that broad view they can be creative and all of these wonderful things can emerge.

The broad view this person sees as so necessary is the balance between anarchy and autocracy. Decentralization was recognized by many as a paradox because it was a process rather than a structure. A principal describes the critical nature of this process:

Someone has to have the idea. And that someone has to contact someone. I mean if it is a teacher the teacher comes to the principal and says, 'I have this idea'. And the principal says, 'It sounds like you've got some good points there. Why don't you sit down with a group of people and come up with a plan and then we'll take the plan downtown and if it fits the general overall policy of the district we'll go for it?' Because even in decentralization there has to be that little thread so that you don't have 35 little empires out there. And that's one of the things that tends to happen in this.

It was well recognized by informants that a school district is not a nice tight entity but that it is spread out over 35 to 40 sites. By its very nature it is subject to communication problems. One principal now recognized that decentralization enabled him to perceive the entire school district as a complex organization. He said he found decentralization to be good for him because he worked well under flexibility and freedom:

But I think I probably wanted too much because I was told, 'You know, even though we are decentralized, you've got to realize you are in a box and on one side of the box are laws and regulations and you keep trying to get outside those rules and regulations. I think you want to create your own unified school district'.

This instruction was handled face-to-face with the superintendent. Another principal explained that you would be notified if you were too far out. All this was conducted with subtlety and sensitivity (see Ouchi, 1981).

If there was a rule it was to make no firm rule. A central office administrator said that for him it was a very statesman-like approach but one that many people could not accept. Thus, a fundamental change from a bureaucracy to a decentralized organization is one of the use of rules. A bureaucracy with its functional rationality usually has rules that apply equally to all regardless of their justness. The rules were loose enough that some principals did not decentralize their schools. Generally, a teacher would not report a principal who had not decentralized. Ultimately, these principals were released of their duties.

The board president said that since rules had largely disappeared, reliance was placed on the good judgment of the persons who were appointed. Freedom of this sort put great stress on administrators in the district who had been accustomed to receiving orders from the central office. Without rules and rank, principals were expected to perform and be willing to take risks. The consequences were not everything that was attempted worked nor did all of the programs succeed. Some mistakes were made. A central office administrator said:

> Because of some of the programs we suffered some. We embarked on this process. And the decision-making was very loose so that people, in fact, could make decisions and we meant for them to make decisions. And sometimes they could be a little bit marginal.

According to Peters and Waterman (1982), allowing marginal decisions is necessary to open a system to develop the trust necessary for risk taking. One administrator said:

> You had to try it and when you try it you have a lot of pitfalls. Some things don't work to save your neck. So you have messes that develop. You have to have the wherewithal to go out and clean it up and go ahead and not get too alarmed if you make a mess here and there.

The fundamental change in rules meant to some a pretty radical departure from traditional practice. While it tested the mettle of administrators, the real risk was to the leader. A principal stated that Berry succeeded because he was able to maintain control.

The conventional organizational chart for a school district is shaped like a pyramid. Typically, the bottom layer consists of the classified employees with teachers, principals, central office, superintendent and board shown in ascending order of power and authority. Students, parents and citizens are normally absent on virtually all school organizational charts. The pyramid chart represents the Weberian ideal of the way in which most organizational leaders think the organization *should* function. Leadership in Riverside Unified School District during the period under study did not consider the pyramidal configuration of power an ideal for organization life. The traditional organizational chart was radically changed.

During the process of decentralization between 1960 and 1971 no organizational chart was published. One principal said, 'You never saw one!' A district administrator said, 'I used to draw an organizational chart for myself sometimes to understand what I was in charge of. I would take it out and look at it from time to time'. One district officer said, 'Oh, well, we didn't do much of that sort of thing then. Not until 1978. Everything had to be in black and white then'. In an effort to find an organizational chart a most cooperative secretary said to us, 'I've checked with everybody. There is no chart before 1970'. The district, 'lacked even an approximation of an organizational chart' (Peters and Waterman, 1982, p. 269).

The 1970 chart is two pieces of paper containing job descriptions but it has no lines or differential heights to represent rank order. It cannot be called a traditional organizational chart, yet it was given to us as an effort to help. 'It was the best the district had to offer', a district administrator said.

Finally, two organizational charts were published in *The Report to the Board of Education on the Organization and Direction of the District*, (Berry, June 1971). One is called, 'The staffing chart which existed prior to full decentralization … '. The chart has roughly four broad hierarchical rankings of positions as shown in Figure 3. Page 17 of the same document pictures the decentralized district organizational chart. The chart is accompanied by the statement:

> The staffing chart which has gradually emerged with decentralization is difficult to show in usual line form. Flow of responsibility, authority, and communication is more complex. The new organization, which is still in development, can be roughly shown as … (Berry, 1971, p. 16).

Figure 3   Predecentralized Organizational Chart RUSD.

THE STAFFING CHART WHICH EXISTED PRIOR TO FULL DECENTRALIZATION WAS ROUGHLY :

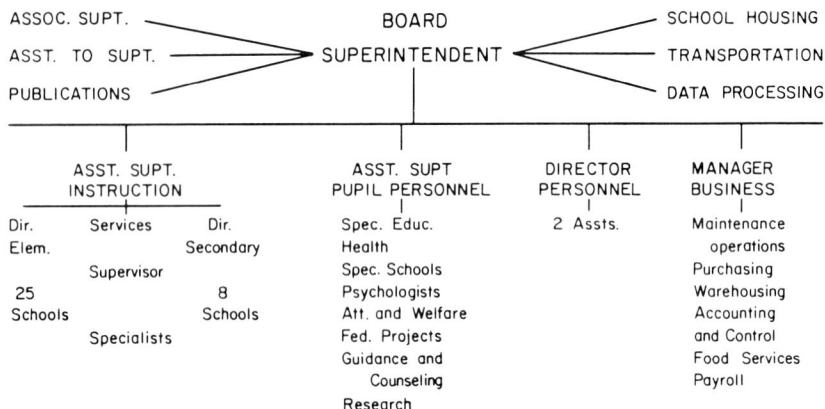

Source: From a Report to the Board of Education on the Organization and Direction of the District, Berry, 1971:16.

What is shown is seven circles, one inside another as shown in Figure 4. These two organizational charts, both published for the first time in 1971, show the difference from centralization to decentralization. The pre-decentralization one is hierarchical in a vertical sense though it is somewhat flat. The decentralized circular chart represents an attempt to reflect the fact that teachers, children and parents are to be involved in decision-making. In its series of seven circles the three representing teacher, children, and parents are in the center. Since decentralization included input from all and placed responsibility on all people in the district, the revised organizational structure and changed roles are represented through a circular chart. Interestingly, nearly every circle has some overlap into another area. For instance, the teacher and principal circles overlap the Division of Instructional Services, the Division of Administrative and Personnel Services, and the Division of Business Services. And all three of the circles respresenting those divisions have a part of the circle respresenting the Superintendent. Authority is shared.

If there is any top to the chart, the circle which is labeled 'community' encompasses the most area and is at the top. This reflects the fact that Berry believed that the school could not accomplish its mandate without the active knowledge and support of its community. In his view, the schools were owned by the community and the

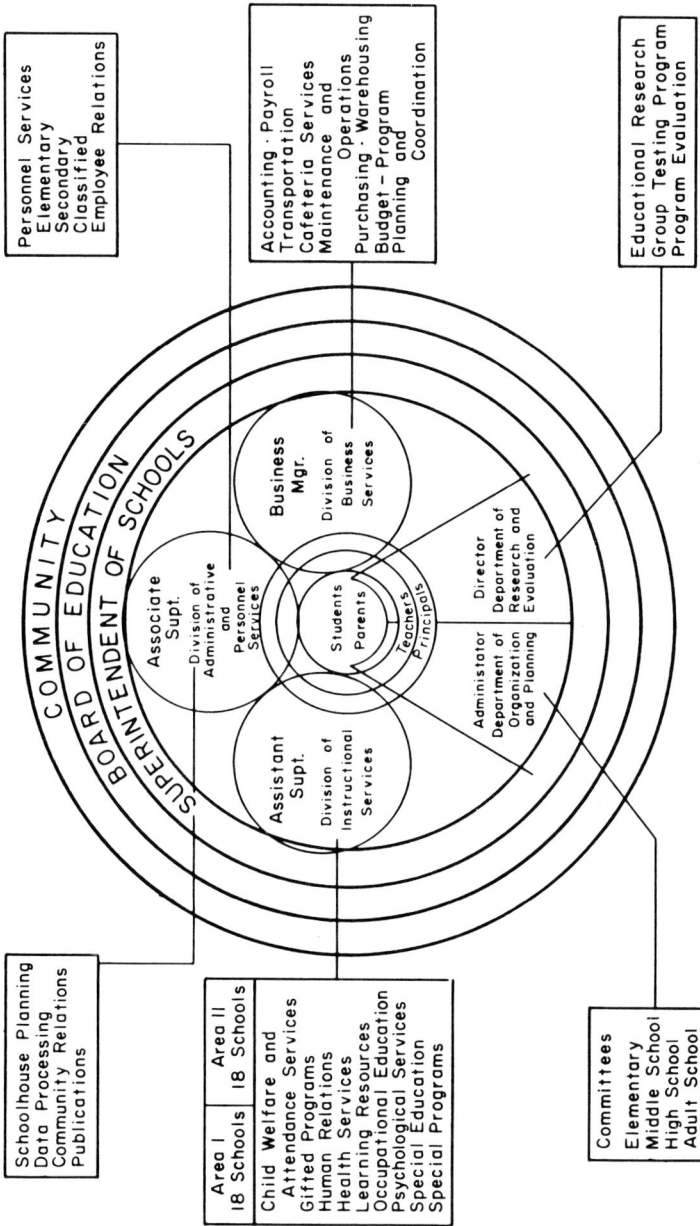

Figure 4   Organizational Chart for Decentralized District RUSD 1971.

Personnel Services
Elementary
Secondary
Classified
Employee Relations

Accounting · Payroll
Transportation
Cafeteria Services
Maintenance and
Operations
Purchasing · Warehousing
Budget – Program
Planning and
Coordination

Educational Research
Group Testing Program
Program Evaluation

Schoolhouse Planning
Data Processing
Community Relations
Publications

Area I 18 Schools	Area II 18 Schools

Child Welfare and
Attendance Services
Gifted Programs
Human Relations
Health Services
Learning Resources
Occupational Education
Psychological Services
Special Education
Special Programs

Committees
Elementary
Middle School
High School
Adult School

COMMUNITY
BOARD OF EDUCATION
SUPERINTENDENT OF SCHOOLS

Associate
Supt.
Division of
Administrative
and
Personnel
Services

Business
Mgr.
Division of
Business
Services

Assistant
Supt.
Division of
Instructional
Services

Students
Parents
Teachers
Principals

Administrator
Department of
Organization
and Planning

Director
Department of
Research and
Evaluation

Source: From a Report to the Board of Education on the Organization and Direction of the District, Berry, 1971:17.

organizational chart clearly pictures that fact since the parent is in the center and the community surrounds the whole. Responsibility and power were spread to all positions. The result was a changed and different decision-making process. That process has been described. For example, a principal tells us about it:

> So we started developing what's called the decision-making process and how it would affect Riverside. We developed a number of models. Where the decisions could be brought into focus by different and varying levels and how each model would affect that particular person.

Mr Berry explained that the experimentation was an effort to see that the person affected would be involved in the decision-making. The effort focused on the fact that changes could begin anywhere in the circles but that much of the action and most of the decisions very largely affected instruction and children, so therefore that level should be pulled into the development of programs. For this reason students and parents are pictured in the center to represent their ownership of the schools.

In the sense that a circle has no top or bottom, the decentralized circular chart has no vertical rankings. Actually, little can be told from this organizational chart about the span of authority or control. One can tell, however, what sort of image this organization wished to project. Riverside Unified School District's round chart would indicate, at the least, that it wished to represent itself as non-traditional and as non-hierarchical. The placement of power and the placement of decisions in varying levels reminds one of the Peters and Waterman story that 'people at Digital don't know who they work for' (1982, p. 319).

Comments by members of the organization about the district confirm that the unorthodoxy in the chart was practiced. One principal said, 'There weren't a lot of charts and graphs so you didn't know what ranked over whom. You didn't have a level'. One administrator explained why the organizational chart was in circles by telling a story about the superintendent, 'I have seen him over and over stand at the board and talk to principals or to others and draw circles. Always circles, never hierarchical charts. He hated hierarchical charts'. A principal said, 'He hated pyramid type things. You know where there's the superintendent on the top and the kids on the bottom. He didn't like that'. There was a perception among the people interviewed that the circular hierarchical chart was an effort to picture and focus the new direction of the district which was to turn all effort to face toward the teacher and the child, to have free communication across as well as up and down levels, and to allow or to encourage decision-making to begin in any of the circles.

An explanation of the district's decision-making process published with the circular organization chart stated:

> An idea or need for a decision can start anywhere within the rings. The very nature of the educational process virtually guarantees that a high percentage of such decision initiations will occur within the teacher-principal center because staff members are largest in number, and they are closest to the children and parents (Berry, 1971, p. 20).

The same document explains that, though the idea may start anywhere, there is a process and if a decision is limited to only one or two levels, the process of participation is killed. Constant efforts were made to contain and to discourage top-down decisions while encouraging bottom-up ones. Decisions in the new structure were to be made by conferring on the issue face-to-face with those most concerned in a manner that recognized the professional integrity and responsibility of each staff member (Berry, 1971, p. 21).

Conferring on the issue face-to-face with those most concerned was not new in Riverside when collective bargaining was implemented by the state in Riverside in 1976. Since it came so near the end of this study we do not have extensive data on it. The data show that collective bargaining was thought to have jeopardized decentralization, but to the degree that decentralized decision-making was already developed in the schools, the two have continued as parallel systems.

The board president remembers its beginnings as a time when flexibility was constrained. A teacher member of the first negotiating team said that collective bargaining had a 'big impact on decentralization. When we got the contract we had to live up to our side and they had to live up to theirs, and there could be no breaking of the rules'. There were uncomfortable times with rigid rules. Teachers and administration worked long and hard over the contract wording and performance clarification. 'We almost had a strike', said a teacher. 'It was at that time', the teacher continued, 'that teachers were no longer given decision-making. It just seemed to go that way and all the schools were saying this'.

Even though there was a period of confrontation and adjustment, ultimately collective bargaining meant only another group getting restructured to protect and to justify its self interest. The district, because of decentralization, was used to permitting decision-making to a variety of groups; the collective bargaining group was another one to be included.

One administrator said that at first the union had requested that the union be the one to select teacher representatives for the cluster meetings. Instead the board approved a policy which stated that no negotiable items would be discussed at cluster. This freed the decentralized decision-making process having to do with mission and instruction to continue unabated. Cluster representatives continued to be selected by each school.

One administrator who may be the most knowledgeable about collective bargaining in the school district said:

> I would like to believe that decentralization occurring prior to the implementation of collective bargaining by the state softened the impact by helping people understand prior to collective bargaining that they could influence the decisions made through the organization rather than collective bargaining. After collective bargaining it may have given the employee two bites of the apple. I would like to believe that it helped to prevent strife. But there is no way to test a preventative measure.

The sides now in retrospect seem to admire each other. The teacher union negotiator spoke about the first administration negotiator by saying, '*He was excellent*. I knew him as a friend. I didn't think he had it in him. But he was *good*. He did his job well'. And at the same time the administration negotiator says:

> All collective bargaining is a centralizing force on a school district. It insists that all employees have the same work hours, rights, training — so school management had input from our side. But bargaining is shared decision-making with a set of people. If the sides are relatively equal, decision-making becomes a set of compromises.

Decentralization in Riverside may have allowed the sides to be equal. To date, Riverside has experienced no strikes. Possibly due to decentralization. No one knows. As the administrator stated, there is no way to test prevention. Unlike the New York City experience reported by Fantini and Gittell (1973) the confrontation did not end in a series of crippling strikes. In Riverside the end seems to be another cooperative inclusion. Collective bargaining became part of the Riverside circular organization chart and decision-making process. Berry wrote:

> *This process is all important, because it results in the necessary attitudinal changes, the vigor, and the leadership that are ultimately necessary for true superior contribution and goal attainment!* (Berry, 1971, p. 11, Note: emphasis in original).

The circular chart and interview data indicate that responsibility for making decisions had changed chairs from *ex-cathedra*. That fundamental change from a centralized structure is reflected in an absence of an organizational chart, an absence of hierarchy, face-to-face conferring on issues, and the assumption of power at the site level.

Goals

The fundamental goal in Riverside was actualizing the mission of the school — instruction of children. Status equalization was the means used to accomplish this function resulting in decentralization. To be rational in many hierarchical organizations is to hold a negative view of persons and consequently to seek to control them, not to release their power (see Hanson, 1979; Ramos, 1981). Rational also usually means that an organization makes detailed and long-term plans and goals. Riverside began its restructuring of a school district organization by modifying those definitions of rational. It concentrated on the quality of means as well as ends, it began with a positive assessment of people, it released power to people, and it did not write detailed long-term goals.

Next there was a redefinition of the Weberian axiom of a hierarchical structure in which each officer is responsible for the actions and decisions of subordinates. First, it had no organizational chart, therefore no ranks. Then it had a leader who was both authoritarian and democratic as the need arose. Berry at one time said, 'I ordered them in'. It was unusual for Riverside. But it was always a possibility. With a positive view of persons the leadership could control when necessary, but it relaxed control the greater part of the time to allow for creativity.

Another tenet of classical organizational theory is that power must be centralized in the chief officer (Hanson, 1979). In this case the chief officer controlled through informal regular communication which aroused confidence in the employees. It was control by the inbuilding of purpose and by challenging teachers and principals to create. This type of positive control, a call to commitment, is ultimately a process of education. We have evidence that the leader used many occasions to instruct. Instead of centralizing power the superintendent was the foremost exponent of sharing power.

In the classical bureaucratic organization formal goals direct events (Hanson, 1979). The goals in Riverside were to teach every child and to make decisions close to the action. The upbeat message was to make things happen that are good for kids. Instead of detailed goals there was a bias for action and the schools became a 'loose network of labs and cubbyholes populated by feverish inventors and dauntless entrepreneurs who let their imaginations fly in all directions' (Peters and Waterman, 1982, p. 14). This experimentation was encouraged in the interest of children. This bias for action allowed everyone a goodly number of mistakes. Formal goals perhaps would have prevented messes and overlap and duplication, but at the expense of creativity and excitement. Riverside teachers were given wide latitude to re-invent the wheel 'because if it was their wheel it stimulated people working with children', a principal said. The role of the leader instead of setting goals was one of 'orchestrator and labeler: taking what can be gotten in the way of action and shaping it — generally after the fact — into lasting commitment to a new direction. In short, he makes meaning' (Peters and Waterman, 1982, p. 75).

With loose goals, with everyone encouraged to take risks and act, with some mistakes, overlap and duplication, the leader still found that 15 years later he was looking at an entirely different organization. There had never been a formal stated goal to make a new organization. There had been an upbeat message, a call to commitment, a maker of meanings, and feverish action (Peters and Waterman, 1982).

'In' and 'Of' Changes

'If we want to change, we fiddle with strategy, or we change structure. Perhaps the time has come to change our ways' (Peters and Waterman, 1982, p. 3). Riverside changed its ways. Joseph B. Giacquinta tells us that there are two basic types of changes which can take place within an organization. One he calls a 'change in organizations' and the other he calls a 'change of organizations', the former involving intentional changes in organizational procedures but not status arrangement and goals and the latter involving intentional changes in statuses and goals (Giacquinta, 1973, p. 180). The changes Giacquinta reports in his review are mostly in the first category because changes 'in' are frequently tried and changes 'of' seldom are. Giacquinta notes the dichotomy and says that 'it may be important in structuring the change process' (Giacquinta, 1973, p. 180).

There were changes in Riverside in both categories. If we define decentralization as a change of organization, the indicants would include social relationships, person-to-person, person-group, and group-group. The change would also be fundamental, comprehensive and internal to people, and the change would take place without mandate. The example of decentralization put into place in Riverside as it was intended and implemented was a change 'of' organization and not simply a change 'in' organization. It changed roles and statuses resulting in changed social relationships. When that happened there was status equalization. With status equality, persons, no longer caught up in where they were on the hierarchy, knew what they should do and did it. After attending to the 'of' change persons could turn to the 'in' or the mission of the organization's functions.

The 'of' changes involved new expectations for and of every role and so basic changes occurred in persons and in their roles. What took place in the case studied in Riverside was a complex resocialization, as Giacquinta (1973) names it, of an entire organization so that internal attitudes, behavior norms, and habit patterns of exclusion in a hierarchical bureaucracy were extinguished and replaced by inclusive ones.

Re-socialization is an important and complex and unusual method of organizational change. One must note that what happened in the study reported in Riverside was internal to the organization and internal to the participants. It was not forced onto the organization from outside political pressures. What we have found and presented is a process which changed people so that they turned to pay attention to what they were doing. It is possible that from the outside they appeared to be doing the same things, but interview data showed that the jobs were done or the roles were filled with the new understandings and new intentions. Case studies of decentralization have normally recorded unsuccessful re-socialization of the participants in the process (Fantini and Gittel, 1973).

Since this study defined decentralization as an 'of' change in status and roles, it focused specifically on the process of those changes. The 'in' organization changes we find may actually only sucessfully flow from decentralization as we have defined it, but this research focused on the process which got to the different social relationships — not changes in activities, curriculum, and program. Woven through the study there are reports of changes in curriculum and program which did flow from the process but they are secondary in this report. The changes we describe are of roles and statuses. It is our thinking that decentralization in a new way occurred in Riverside and as a result there were changes in: relationships, structure, program, and curriculum. Giacquinta (1973) would call the first two 'of' and the last two 'in' organizational changes.

Relationships changed in terms of participation. There were specific changes in the valuing of participation, the reasons for participating, in who participated in what, and how it was done. Citizens, parents, students, teachers, and principals were invited, expected, challenged, and at times forced to take part in decision-making as peers. They were invited in on the ground floor of policy formation and continued to be a part of the process as adjustment and readjustment took place.

Other case studies (Gittell and Hollander, 1968; Steinberg, 1977) report where participation may have increased but the quality of participation remained negative. LaNoue and Smith (1973) reported that there had not been a substantial increase in parental power. Steinberg (1974) found that parents were still unable to influence the kind of schooling offered and Lawson *et al.* (1973) reported only little-to-some progress made on decentralization goals in Los Angeles as a result of increased parental input. Steinberg (1977, p. 121) reported that increased participation without access to the decision-making process can actually increase alienation because 'participation leads to positive attachment only to those institutions in which the participants feel some measure of control and influence'. Riverside decentralization data show that most parents and professionals had an increased affection for the organization. It was said by a parent, 'We became one big family'. Principals said, 'We never wanted to leave here'. Administrators who had left for better jobs experienced disillusionment when they did not find decentralization elsewhere. Parents, non-professionals, and professionals experienced some measure of control and influence. For example, parents, custodians and cafeteria workers were included in some budget sessions. Principals made decisions on budgeting which were once only central office prerogatives. Teachers made curriculum decisions which were once far above them. While other case studies are a litany of efforts of the organization to exclude as many as possible from the decision process (Gittell 1967 a and b), Riverside was a case of an organization making every effort to get all the meaningful participation it could.

A second specific 'of' change in Riverside was that the structure moved from pyramid to circular. This change was reflected in a honed down central office, a changed role for the central office, and the way in which the office communicated. Control moved out of the central office as it became a service organization to facilitate programs which parents, teachers, and students helped design. Communication was by instruction, not edict. Persons in the office were expected to become emissaries for the process and instructors through example by consensus decision-making. Administrators said that, as people learned the new structure, 'jobs became easier, more fun and enjoyable'. Increased sensitivity was reported; for example, professionals drew the parents in. Other decentralization case studies generally report the professional in the school system as resisting decentralization (Fantini and Gittell, 1973; Steinberg, 1974; Peterson, 1976).

A third specific change in Riverside was an 'in' change as programs changed when they moved from under the control of the central office out to the sites. Status and roles changed about who made the decisions but there was also an explosion of creativity. Every school in the district was encouraged to run its own program to meet the needs of the particular group of students and parents it served. Principals took great pride in developing their own schools. As a consequence some schools responded by becoming unstructured, non-graded laboratories teeming with experiments. That option was open to all schools in the district. There was, as one administrator said, 'fractured curriculum!'.

Decentralization in Riverside was district-wide. Even when New York and Detroit were forced to decentralize their total system, they did not move to decentralize program decision-making to each school site (Fantini and Gittel, 1973; LaNoue and Smith, 1973). Decentralization in those cities involved community school boards being elected and being given some discretion. Riverside is a demonstration of district-wide site power.

The issue of who remained in control of the school district is also well illustrated. For example, the final explosion over decentralization in New York City occurred when one site, IS 201, exercised site discretion by transferring 19 professionals to the central office without the concurrence of the superintendent (LaNoue and Smith, 1973). Riverside's leadership was never questioned. Such occasion as teachers sitting in on the hiring of their principals, deciding what grade they would teach, and agreeing on the number of children in each class were common, but they were not challenges to the leader. The radical structural change which occurred was site power and independence while the central office became a professional support system. This is in contrast to other cases of decentralization in which the central office retained its decision-making function for the school district (see Steinberg, 1977).

There was a fourth change: productivity increased, morale shot up, commitment deepened, and with that change 200 to 250 curriculum parts were developed. These are the 'in' type of change in school organizations which the literature reports extensively. They are important to the student but we think the Riverside decentralization case shows us that they are more meaningful and enjoyable when 'of' changes have taken place first. Riverside experienced significant curriculum developments. The Riverside Instruction Support System (RISS), which was a method of tracking each student, was a specific change put into place system-wide. It resulted from and resulted in an intense focus on each student enabling the teacher to know exactly what was mastered and what needed to be taught next. One teacher called it a 'meticulous record of each child's skills'. Even one such specific change put into place in a district is an indication of dedication to the mission of the school. However, there were at least two other such major 'in' changes in the district curriculum as a result of decentralization.

A series of 20 books, 1–16, 17a and b, and 18a and b, The SAM Series (Vergeront, 1971), was developed by teachers for kindergarten children to use at home with their parents. These were written by a teacher and then published and produced by the district. The books are designed for second semester kindergarten children dealing with pre-reading and reading skills. The books contain adult and children's pages which provide a language approach to reading. The word SAM is drawn from the main character, Sam the lion. Numerous other districts judged them good enough to purchase. This creative curriculum idea of each child having his own book to share at home attests to the school personnel valuing the child, the home, and the books. Secondly, district personnel helped write, pilot and publish a new social studies curriculum which incorporated a section, *The Third Culture*, in the text. This curricular piece was a bold effort to be inclusive of all cultures and allowed the teacher to address the affective domain as basic skills were taught. These curriculum changes were said by teachers to have been possible only 'because we were decentralized and the district was going to support teachers who do creative things'. These curriculum changes were consistent with the kinds of relationships and structural changes which happened in the district. The 'in' changes in curriculum reflected the intent for inclusiveness. Inclusion of a third culture in curriculum showed valuing of difference. Inclusion of the affective domain reflected valuing of the whole person. The curriculum changes were a part of the process and not implemented programs as gimmicky new ideas or innovations for the sake of change. As the 'of' changes in the organization were meaningful, so were the 'in' changes.

In summary, decentralization in Riverside encompassed the two types of changes which Joseph Giacquinta (1973) notes as a dichotomy. The 'of' changes which changed status and roles were in the areas of participation and structure which resulted in 'resocialization' from exclusion to inclusion. The 'in' changes in the areas of program and curriculum were numerous and, more importantly, were an integral part of a whole process.

Contrasts

Riverside decentralization differed from all others in the literature because its impetus to decentralize was not from the external environment and was not forced onto the system. LaNoue and Smith (1973) state that in the cities they studied other than St Louis, the 'community control movement has required either a major center of political opposition to the school establishment or an issue that it can use to mount an attack' (p. 56). The literature also shows that schools did not even administratively decentralize without pressure from the outside environment (Pilo, 1974; Peterson, 1976).

The events that pushed schools toward decentralization vary from city to city and the literature is either preoccupied with those events, the variation in the response to those events, or to the outcomes (Steinberg, 1977). In New York numerous studies found that it was dissatisfaction with the schools and their failure to integrate. In Detroit school officials were pressured on integration as Blacks supported decentralization there, and in Los Angeles LaNoue and Smith (1973) state it was Black and Mexican-American groups who raised the issue. In each city the issues triggering the call for decentralization were thought to include the failure to educate and the failure to integrate classrooms. Decentralization preceded desegregation in Riverside so that factor was not a triggering issue in Riverside's decentralization (Hendrick, 1968; Berry, 1971; Gerard and Miller, 1975). Riverside differs in this respect, we think, from every other city.

Internal factors contributing to the demand to decentralize are listed in the literature as unresponsive central office personnel, unresponsive school site officials, massive student failure, and teachers who felt powerless. External factors triggering demands were thought to be: parents who demanded participation, the size of the city, failed integration promises, federal laws, foundations, trends in the culture, and the mass media (Howe, 1968; Fantini and Gittell, 1973; Tyack, 1974; Moore, 1975).

We found that while Riverside had a centralized system and some unresponsive school officials, there was no major political opposition to the school establishment (Hendrick, 1968). Parents were not demanding decentralization. It was also a smaller city than those LaNoue and Smith (1973) found to be correlated to the demand for decentralization. There were no foundations pushing for decentralization with manpower and money as was the case with the Ford Foundation in New York City and Danforth Foundation in St Louis. Federal funds were thought to be the trigger in some places, but in Riverside decentralization as a concept and as a goal for the district preceded the availability of federal Title I and III Elementary and Secondary Education Act (ESEA) funds.

In other cities pressures external to the system triggered and/or forced whatever decentralization (rhetoric or actual) took place. Pilo (1974) said that 'both New York City and Detroit decentralized their school systems only after the state legislature . . . required them to do so' (p. 5). Even after citizens turned to lawmakers to impose the right of citizen paticipation in decision-making onto their schools, the transferring of authority conformed only to the point of some administrative decentralization. The reaction of the system and of many professionals in it was to thwart the demands. Peterson's (1976) statement about Chicago would stand for most cities reported in this literature. 'Political decentralization was delayed and undermined by a staff committed to maximizing its own discretionary powers' (p. 254). He claims that even the authority was altered only marginally.

The literature on decentralization reveals that in four cities — New York, Detroit, St Louis and Los Angeles — the response of the school boards to decentralization demands was a symbolic institutional agreement in principle.

> When push came to shove for decentralizing systemwide as it did in every city . . . the boards split into racial and ideological factions. No board fully supported community control and most sought to modify or defeat legislation introduced at the state level (LaNoue and Smith, 1973, p. 226).

The literature also makes it clear that the leadership for decentralization has not come from professional educators (LaNoue and Smith,
1973).

Riverside seems to be a different case. We found in Riverside that
it was not factors external to the system which raised the issue and
were the impetus to decentralization. It was a single organizational
leader who set out to improve his school organization and in the
process instituted decentralization. So while other systems had decentralization imposed upon them from the outside, the intention and the
origination of decentralization in Riverside were different. It was
sought as a positive change 'of' the organization by the leader. The
focus was on the school mission with an intent to enrich relationships
between people, thus insuring the means as much primacy as the ends.

Secondly, the effort to decentralize the organization began at the
top of the organization instead of at the bottom. When the four cities
mentioned before responded to decentralization demands, they did so
by trying to institute changes at the bottom of the organization by the
introduction of experiments. Such efforts are symbolic since they are
far from, and do not necessarily touch, the central office hierarchy.
They are bits of decentralization safely contained off to one side while
the central office 'mossbacks' (LaNoue and Smith, 1973) may proceed
unaffected. It is reported that central office people will allow changes
far from them in program, curriculum and activities, that is, 'in'
things.

In Riverside decentralization struck early at the heart of the structure, philosophy, behavior and beliefs of the central office. Beginning
to decentralize from the central office out we think was a factor
contributing to success. As Pilo (1974) says, even when decentralization was forced the response corresponded to administrative decentralization and his thinking is that even that would not have happened
without political pressure. Fantini and Gittell (1973), Ornstein (1974)
and Steinberg (1977) found that the professionals, usually meaning
central office personnel, retained power in any case. This meant that
the movement which represented 'an effort by powerless groups to
become a part of the system and ... to make it responsive to their
needs [failed]' (Fantiti and Gittell, 1973, p. 7). They say such groups
were effectively excluded from sharing a responsibility in the alloca-
tion of societal resources. Starting decentralization at the top in Riverside indicated that changes were not to be cosmetic but real and
comprehensive.

The third difference was that the process of decentralizing was by instruction and not by order. LaNoue and Smith (1973) mention a St Louis superintendent who 'viewed the office of Superintendent as the master teacher of the school system and the city as his classroom' (p. 40). However, he sometimes viewed his fellow men with an 'all-encompassing cynicism', and the policy of 'anticipatory gradualism' of his board was challenged by some responsible persons as 'paternalistic and colonial'. LaNoue and Smith (1973) further say that the school leadership in St Louis 'has a formidable ability to resist sharing power' (p. 58).

Decentralization in Riverside which used instruction as a tool did not contain the element of exclusive cynicism. It was called 'instruction by example' and as such it mitigated against cynicism. In Riverside instruction resulted in changes in participation, structure, program and curriculum while the top-down ordered decentralization in New York and Detroit was difficult to exercise because of internal constraints or resistance.

Decentralization in Riverside was not instruction of but was along side of. Peters and Waterman (1982) say that we are short on practical design ideas for what they call the software or soft 'Ss'. Those are such things as organizational style and shared values. The leader in Riverside seemed to sense that and he learned along with the staff. He experimented over and over again. His first effort at cluster organization did not gel, and flexible scheduling in one school nearly destroyed it. But throughout the whole effort there was a sense of the inherent worth of what they were after. They were not guilty of what Herzberg (1966) observed of most American managers when he said, 'Managers don't love the product. In fact, they are defensive about it' (from Peters and Waterman, 1982, p. 36). Some of the Riverside participants were so intent on the product, instruction, that they valued experimentation and were not aghast at mistakes and expected to go out and clean them up. To value experimentation means freedom from a home run mentality. Riverside school personnel made mistakes because they risked activity, 'having escaped from the tyranny of reason' (Peters and Waterman, 1982, p. 40). The decentralization in Riverside took place in what the two authors would call an excellent company because it was a learning organization. So the Riverside case study was different in its use of and valuing instruction.

Decentralization in Riverside was called by several, an 'evolution in the system'. Peters and Waterman (1982) say excellent companies have seeded evolution. One principal referred to Berry as 'planting seeds'. Evolution and the growth of seeds are time consuming. Decentralization in Riverside took the 18 years Berry remained in the district. There is evidence in his personal files that he was still working on it on the morning of the day he retired. What Riverside shows us is that the leader must experiment over time as part of his and others' learning. Other decentralization case studies have looked at the system for shorter periods of time. LaNoue and Smith (1973) examined five cities for three years. Riverside points out to us that internal decentralization of the 'of' type required almost two decades.

In summary we have pointed to some differences in the Riverside case study from others in the literature. It was internal, not symbolic, but actual; it came from a professional educator; it began at the top of the organization; its instrument was instruction; and it took a long time. There were similarities to other systems, too. Chief among them was the fact that some parents and some professionals resisted the process from the first and on to the end.

As a final word we should say that this study covered only the years 1960 to 1978. Therefore, it does not examine in detail what happened after Berry retired. Informants mentioned that there had been recentralization to a degree. Others said decentralization had not been lost. What happened after 1978 remains to be determined.

Chapter 5

The Control of Information and Leadership Success

The literature on organization life demonstrates that conflict is ever present and institutions are constantly challenged by internal or external forces. In examining the superintendent's office, the 'controversy seems to be continuous'. Blumberg (1985, p. 20) claims that it is 'directly related to the issue of power or perceived power. If a superintendent has (or is seen as having) power, that power has potential for affecting a community's value system, its pocketbook, and the welfare of its children — among a host of other things'. The same author quotes Knezevich (1975) who noted that the superintendency is 'a position born of conflict' (p. 372).

In looking at school organizations during the decentralization movement in the late 1960s and early 1970s, one of the most pervasive descriptions is that of conflict between organizational levels. For example, Rogers (1968), in his study of the New York schools, found that when the community rebels against the central office and internal battles between different units of the central office take place, the bureaucracy cannot function without its orders flowing from the top to the bottom, yet rational decision-making and implementation is impossible at the top of the bureaucratic structure. (See Borman and Spring, 1984, p. 82.)

Borman and Spring (1984) explain that introducing reform into the New York Public Schools was difficult because it originated externally. In contrast, the superintendent in the present study implemented internal change which proved to be better because external change frees people to oppose. It frees people to rebel against the executive. As McGivney and Haught (1972) reported in their study, 'it is precisely during the periods of heavy criticism that the central office staff seeks to maximize its control' (Borman and Spring, 1984, p. 88).

In contrast to conflict regarding the central office, there is also conflict relative to principals. Morris, *et al.* (1984) in their study of 24 Chicago principals found that principals exercise discretionary insubordination of disobedience to rules. This is done in several ways: unwillingness to make any major impact on the bureaucratic hierarchy; holding a gentleman's agreement of disobedience among themselves; planning a delinquency of deadlines; and disobeying an instruction by following it literally.

Another aspect to conflict in school districts is that related to hostility toward administrators residing in board–superintendent relationships. Blumberg (1985) reports that as early as 1891, John Clark, superintendent of Flushing, New York resigned over the question of 'Who is running things?' (p. 22). The fundamental conflict resides in the issue of 'executive power and management prerogatives' (p. 26). Sizemore (1981) describes this type of conflict in her study of the Washington, DC schools. The struggle took place among school board members representing different priorities and against the superintendent and other school officials. Sizemore concluded that the basis of the conflict was due to the avoidance of the District of Columbia Board of Education to formulate a conceptualization of decentralization or to chart a strategy for the school system to adopt. Instead, it accumulated a set of precedents in the form of contracts with the various boards as they assumed temporary power. These individual boards represented groups which fluctuated in their support for the superintendent and other school officials.

The superintendent, on the other hand, methodically prepared plans and strategies for decentralizing the school district. Arnez (1981) describes the conflict which ultimately led to the failure to decentralize the Washington, DC schools. Arnez identified ideological differences, role confusion, and racial and sexual conflicts which interfered in the process. The author concludes:

> Therefore, the decisions regarding Black push for local control of schools, the desire to raise the academic level of students, and the need to equalize expenditures of the Washington schools' system resources in schools located in rich and poor areas of the District, were more political than they were educational. The decisions were those that extended the power and control of the majority on the school board rather than facilitated the learning and development of the student population (p. 427).

Blumberg (1985) in his study of the superintendent living with conflict, details the areas of conflict confronted by the superintendent. He concludes:

> The highly political and conflictual world of the superintendent appears to be randomized, irrational, and uncontrollable. To the extent that a superintendent can understand the complex web of events that he confronts and subject that understanding to his own intellectual analysis, he is better able to order the world, control it, and maintain his own sense of self (p. 57).

> Balancing the two demands, one for system management and the other for educational leadership, presents the superintendent with a daily dilemma, whether or not he is conscious of it (p. 209).

> Being responsible for the behavior of others but not being able to exert direct control over that behavior, then, constitutes a real dilemma for superintendents who care, as most of them do, about what happens in the schools (p. 213).

As can be seen from the examples presented above, conflict is lodged between organizational levels and is related to power and control. During Mr Berry's tenure the potential for conflict was always present, but in contrast to the superintendents reported in other decentralizing districts, he controlled information in a systematic way to avoid or to benefit from conflict. The reader is reminded that the superintendent of the Riverside Unified School did not begin with the purpose to decentralize. Rather, he began with the purpose to reform the organization and improve the academic achievement of all children and the work performance of the institution's members. As described in Chapter 3, the process which evolved resulted in a decentralized structured organization. Levy (1966) states that:

> the major structure of decentralization in the last analysis is always the actual if not ideal autonomy of the nearly self-sufficient units within the sphere of their self-sufficiency. Within this sphere the members are almost always left entirely to their own devices. Interference with individual members may be spectacular, but general behavior is centralized only slightly in terms of ... other large-scale units (p. 100).

He further elaborates that:

> high general centralization and low decentralization has not
> been a possible basis for stable large-scale societies until re-
> latively high levels of modernization obtained in communica-
> tions as well as other respects (p. 310).

The consequences from decentralization or centralization are succinct-
ly presented in the following passage:

> The more highly interdependent the various elements in a
> relatively modernized society becomes, the less any adminis-
> trative ingenuity in the form of decentralization of power can
> substitute for detailed adequate knowledge about the areas in
> which decisions must be made. If ... stability is to be main-
> tained, limitation of the amount of actual planning will be a
> function of the level of adequate knowledge available rather
> than of preference for or against planning (p. 489).

The administration of the Riverside Schools by Mr Berry was conducted with one major goal in mind: the improvement of children's
academic achievement. All other purposes for organizational participation were secondary. Associated with the major goal was the realization that children learn in schools through the intentional efforts of
school personnel. This intentional behavior was assumed to be in the
interest of the child, but in order to insure that this was so, the
superintendent used direct and specific means. The major way he used
to maintain stability while he reformed and decentralized the school
district was through the control of information.

Why is it important to examine how information is controlled by
the chief executive of a school district? As Glasman (1986) points out
in his examination of evaluation processes leaders must come to realize that there is a time between the 'moment school leaders begin to
process information and the moment they begin to develop a rationale
for a potential decision or action' (p. 160). This moment is when they
begin to process information intentionally. Their success and/or judgment is probably dependent on stretching or controlling the distance
or 'territory' between the two points.

Glasman explains this process when he describes the activity of
rendering a judgment. 'Judgment rendering begins when (administrators) first process information on the basis of some definite intention
on their part, and ends when (they) first develop a rationale for a
definite decision' (p. 166).

In the present case, one of the issues viewed as primary to the superintendent, Mr Berry, was the differentiated school performance in standardized tests. He sought information which convinced him that he had two choices: intensify the hierarchy or soften it. As was reported in Chapter 3, he chose to soften the hierarchy by decentralizing the school district. He used the means of controlling information to reduce conflict.

The Control of Information with the Board

The first means for the control of information was with the school board. He did this in two different ways. One was formal and the other informal. Informally, he met with the president of the board before the formal meeting. All other board members were informed about this meeting, but they did not attend. Other informal meetings took place, primarily in the evenings, but sometimes for lunch and sometimes in his office. As Mr Berry put it, 'the key was to get the job done'.

Formally, he shared information with the board. He referred to it as not playing games or trying to generate individual votes. His intent was to work with the board in total, without excluding anyone, thus avoiding fragmentation. All board members received the exact same information. The process by which school board members were informed was the following. He presented an idea to the members, compiled their views on the idea, prepared graphs, pictures, whatever to illustrate or emphasize the idea, gathered additional views and perspectives, prepared a narrative dealing with the idea, sent the narrative to school board members prior to the meeting with him, met with the board and reviewed the narrative, determined if the idea had been accepted, revised as necessary and if it was ready to be discarded or to be presented at a formal board meeting as board and school district business. If it was ready to be presented formally, the process just described above was conducted with all relevant groups.

In an effort to avoid conflict over issues, the superintendent provided opportunities for the board to react. If comments were invited, they were accepted. Mr Berry described the process this way:

> When you are asking people to express what they feel, take a lesser role. In a formal setting, however, the superintendent takes the lead. Ultimately, the board decides. Don't make your judgment according to where the board members are.

> Anticipate what board members need to know in order to carry out their jobs. Help them along with their total knowledge.

The processes by which Mr Berry kept the Board informed served to avoid conflict between them. It also served to specify the roles each was playing. Guthrie and Reed (1986) after studying school board and superintendent relations concluded with Gee and Sperry (1978) that:

> the statutory protection of the job tends to be minimal, with employment conditions dictated primarily by contractual agreement negotiated between the Board and the superintendent. Thus, the superintendent who by law may have the right to exercise a good deal of initiative may in fact not be able to survive if the Board is determined to restrict or scrutinize his or her every move. Conversely, the superintendent who has little or no statutory authority ascribed to his or her position directly may exercise great liberty and initiative if the Board is willing to place faith and trust in the superintendent's judgement and give sustained approval to the administrator's ideas, suggestions, and actions' (Gee and Sperry, 1978, p. 58).

Mr Berry had established a relationship with the board in which trust and faith in him permitted wide latitude in initiating district acts.

The Control of Information with Principals

As was presented at the outset, conflict in school organizations can also arise among principals or principals against the superintendent. Ziegler *et al.* (1985) report that the greater amount of conflict for superintendents arises within the organization with other school administrators and personnel. Mr Berry in an attempt to improve education dealt with principals by transforming their offices and schools into autonomous units with their own budgets. The principals in conjunction with teachers developed their school plans and bid for budgetary resources. Conflict sometimes arose, but in general structuring the school sites in this fashion instead served to accentuate challenge.

Kanter (1983) claims that the 'organizational power tools consist of supplies of three "basic commodities" that can be invested in action: *information* (data, technical knowledge, political intelligence and expertise) *resources* (funds, materials, space and time); and *sup-*

port (endorsement, backing, approval and legitimacy)' (p. 159). Whereas Mr Berry could confine his exchange of the three basic commodities to information with the school board, in his relationship with the principals, he invested all three, but information retained its primacy. The budget document and school program proposals are two examples. This information was processed between the superintendent and principals in two major ways. One was through informal and formal meetings between Mr Berry and the individual principal and the other was through formal meetings between Mr Berry and all of the principals as a group. Many times the principals were broken up by levels, so that senior high school, junior high school and elementary principals met separately.

The data collected for the analysis of the decentralization process indicates that intense meetings were held for a period of about six months, before the principals were mandated into the decentralization structure. Included in that mandate was the provision that the superintendent could rotate principals among the schools. In general, this was intended to revitalize and to provide for growth for the principals. The potential for the use of rotation as either a reward or sanction, however, was fully appreciated by the principals. The data also show that it was this group which was most resistant to the decentralization structure.

Several reasons appear plausible. First, prior to the decentralization efforts, the principals had direct exclusive access to the superintendent. As decentralization evolved, principals' exclusive access to the superintendent was tempered by the inclusive access of all other groups. Second, prior to the decentralization efforts, principals sought support and aid from the central office which, in turn, resulted in directions to teachers. After decentralization, equal participation and cooperation between central office and teachers diminished the superordinate role of the principals. Third, the cluster system served to insure that information relayed to school personnel was uniform. The superintendent, directly relaying information to each cluster, insured that all levels of the school system received identical information. In this manner, no group could assume primacy due to its possession of exclusive information. Equally important was the fact that since the superintendent controlled the information for the entire school district, his role as the leader could not be disputed.

There is another aspect to the control of information as it relates to principals and the way it was structured by Mr Berry in order to avoid conflict. When information had to be relayed in the decentralized structure, the information had to be general and applicable to the general mission or function of the school organization, rather than to personal or group benefit. This aspect had two important results. One was that there was reinforcement of the basic mission or function of the school and the other was the style by which this function was to be performed, through cooperation and openness.

In sum, the school budget and the school program proposals became the two major information documents which directed the performance of the principals and their school staff. The structure and the processes by which the budget and programs were implemented served to instruct the principals and other school members how to participate.

The Control of Information with the Central Office

Conflict may also arise among the central office staff or between the central office staff and other members of the school organization including the superintendent. As was detailed in Chapter 3 Mr Berry reduced the central office considerably and changed the role of many of the assignments. These changes contributed to a reduction of conflict, but the importance of the control of information might be best illustrated with this group.

First, as McGivney and Haught (1972) found in their study of the central office, the officials in the central office perceive as one of their major roles the control of conflict and the control of information. Wresting these traditional functions from this office required dramatic acts. The most dramatic being the institutionalization of a research and evaluation unit. Another act was to have the position of an intellectual which coordinated staff data analysis, presentation and theoretical aspects. This position is explained in Chapter 3 as the role of the thinker and creator. The third means was to create a staff position which served to collect, compile and process particular types of current, local and social information such as newspaper releases.

Glasman (1986) charts the development of school evaluation units. He reports that 'by the 1970s, evaluation in education had come to include all of its earlier meanings. [That is], it was the measurement of individual differences, in relation to the achievement of curricular objectives, and its function was to provide information to decision makers' (p. 11). He further claims that 'there is no evidence' that systematized evaluation is institutionalized in the internal organizational structure of local school districts. By institutionalization he means that the unit is 'an integral part of the organization core technology' (Perrow, 1965, pp. 38–39).

The importance of the institutionalization of the research and evaluation unit is an act which can be appreciated when it is realized that Mr Berry had it institutionalized directly under him. The director of the research and evaluation reported directly to him. The institutionalization of this particular unit also meant that other central office characteristics were affected. For example, a public relations office or officer to handle community responsibilities was absent. The superintendent explained, 'Some districts have a community relations specialist. I do not think it is the best system. You have to institute a whole system of communication. All educators have to accept the responsibility for communicating'.

Another characteristic that was affected was that the central office had to engage in providing services associated with children and instruction and not with controlling conflict and/or information. Conflict and information were controlled by the superintendent through the office of research and evaluation. This office was responsible for collecting all student evaluation information for the school district, for the analysis of all testing, student discipline and student services, and for reporting to the board, the press and the community the district's progress in these areas. This office directly under the superintendent insured that the director reported directly to the superintendent, rather than to an associate superintendent of instruction as is common in other school districts.

The superintendent explains the importance of this office in this manner:

> How do you report test data to the public? First, you have a research office which prepares all of the appropriate materials. You might want to describe the process, or present the results. In any case, the intent is to move the district to better achievement. What is needed, then, are facts, e.g., standardized test scores and results. Your issue is that schools differ in their scores. How do you handle this material effectively? You need to know what is needed to promote higher achievement and close the discrepancy between schools. There are a series of steps to take in order to determine the reasons for the discrepancies. The research office assumes the responsibility for reporting these differences and reasons. As a school district you cannot tolerate a wide gap between high and low scoring schools. The research and evaluation office is directly responsible to the superintendent. Objectivity in presenting facts,

reasons and projected actions is imperative. The job done properly emanates from here.

The institutionalization of a school district's data and information collecting and processing unit resulted in school site comparability and accountability. But more importantly, the superintendent assumed responsibility for the school district's performance.

Because evaluation is expensive, public acceptance of this office is important. In the case of Mr Berry, the evaluation process itself was classy and consistent with the paper work carefully prepared. In the transmisson of information he provided explanations for results openly. The staff developed the explanations and laid out the projected acts to improve the weak and negative aspects. This reporting of information was not haphazard nor covered up. When a staff member erred, the error was cleared up. The belief was that most communities provide for mistakes, but not repeated ones. The support services were utilized for the purpose of improving programs.

The assignments of the intellectual to process information for the superintendent meant that central office groups could not assume the role such those described by McGivney and Haught (1972) as they sought to 'control the situation'. The intellectual whose role remained ambiguous to other school personnel performed the function of collecting theoretical information and explanation, developing pragmatic and practical explanations for the superintendent, board and community, and coordinated activities with the academic community. In performing the above tasks, he assumed the public roles of advocate and adversary, but the underlying intent was to provide for the superintendent the best theoretical and analytical information available.

The institutionalization of this post is remarkable in its capacity to reduce the effectiveness of the study groups normally associated with school district business. Central office staff could no longer control school affairs, the image of the school district, or the superintendent and board. (See Gittell, 1967b, for a description of the power the New City central office staff retained.) With one person as an intellectual, directly responsible to the superintendent, the school district leader could wrest the power from the central office staff and lodge it in a more manageable unit. Because the role was unfamiliar and suspicious, the intellectual was ineffective in marshalling power for himself against the superintendent. Instead, the logical relationship between the two was more of a leader and his right-hand assistant.

The third point of information control was the staff position Mr

Berry created. This was the assignment of a former female teacher to collect and process local information related to education, to sniff pockets of organizational rife or unrest, and to alert the superintendent about sudden local and organizational events. This person was not Mr Berry's secretary. His secretary fulfilled the traditional secretarial functions. This new position was created specifically to place local information directly under the superintendent's control.

The Control of Information with Teachers

The teacher–superintendent relationship is usually considered a collective bargaining issue. Mr Berry considered it an issue which insured that the mission of the school institution was fulfilled. He perceived teachers as the critical actors in the improvement of the organization. In regards to the control of information, Mr Berry instituted the cluster system whereby teachers could have direct access to him and his office. A public forum was provided where teachers played as critical and active a role as principals and central office personnel. Because teachers were involved, classroom and instruction as important issues could not be avoided. The inclusion of teachers as members of this group, insured that principals and central office administrators could not make demands upon them without their prior participation. Thus, the importance of the cluster system as an information vehicle cannot be underestimated. In sum, teachers, were informed directly by the superintendent regarding issues specifically related to their tasks in a uniform and systematic manner. Because the information was focused on instruction and children the importance of teaching and the work they were engaged in was heightened.

The Control of Information with the Community

Thus far, the control of information has been presented in relation to the internal management of the school district. The superintendent faces another group which must be kept informed. This group is the community. There is information to be reported to different groups. The groups may be members of the school district; they may be outside the school district. As Mr Berry often said, 'It is important to let the community know from the superintendent first what the district is doing and what it intends to do, rather than have it discovered'. Information will always have to be relayed to the district, to the board, and to the press. A school district cannot function without providing a means for the public to be informed about its work in some systematic way. One of Mr Berry's general practices was to report and show what had been accomplished, for example, test results, performance levels of certain activities or presentation of special events. Regular board reports with staff prepared details provided some of this information. Both the bad and good were reported with the belief that it was better to report the negative than have it discovered. If this process appears terribly inefficient, it should be noted that it is this structure which provides for the relay of information from the bottom up (for example, parents may complain to the board) and down from the top (for example, the board reports).

The superintendent's reports supported through facts and details usually dealt with significant district functions. In order to be effective in relaying information, the superintendent remained informed claiming that about one-quarter of his time was spent in this task. The other three-quarters of his time was in the operation of the system. Progress reports relayed to the public were designed to inform the public about the way he spent his time and about the priorities of the district. In the preparation of the narrative for these groups, the writing was formal for general audience use, but specific and professional using plain and precise language.

The important aspect to staying informed was that it permitted the leader to deal with difficulties before they became public issues. Maintaining that all decision making consider the benefit of students before adult 'interest insured that certain types of conflict were avoided.

The staff assistant aided Mr Berry in coordinating written, media, and miscellaneous information. Materials which provided basic school district information included the *Riverside Board of Education Policy Handbook*. The superintendent was responsible for monitoring the updating of the manual and adherence to it in the implementation of its rules and regulations. Discipline and personnel matters were the most common areas covered.

There were several public means by which Mr Berry kept the community informed. One was through the local press. He worked in a close relationship with the press to help it be better informed than it would be and as often as possible with positive stories. The strategy was to have the superintendent's office informed about what the press was reporting in regard to educational issues. They kept the clippings and regularly analyzed the contents in order to determine the types of stories to relay; Mr Berry put it this way: 'The press will take the stories we give them if we do it the right way. You make certain that they can get these stories easily and comfortably. You have to know what the good story is'.

Mr Berry described the process he used:

> Provide background information when you are developing a major story like the national test program. Sit down with them or have your research people do it and cover the topic in depth and get their agreement to hold the story until the board meeting is over. This way, the story that is released is yours and not the reporters.

Mr Berry explained why the development of good press relations is important:

> You need to realize that the media are business. Newspapers, television, reporters ... they are not designed, they don't exist to tell your story, to put your job out front. They are businesses to generate readership which, in turn, generates ads and impacts on the community. It is big, powerful business.

Another means to relay information was through television. Mr Berry described it this way:

> Now that TV has the technical sophistication, they can come to you and put you on TV personally that day if you have something going on that interests them and that is a good story and can provide good pictures. They can walk into your office and plug the stuff in and you are on! Like that! Bang! And maybe, you don't get any warning or not more than an hour's warning. They do not want you to have more than an hour's warning. They do not want you to be ready. In my case, it happened to the president of our board, who happened to be a remarkably, sophisticated, intelligent guy, being inter-viewed on a curb outside his office. It can happen that fast if you are willing ... And I would suggest that most of the time you had better be willing. You can't turn them off because they'll say, 'Well Superintendent so and so or Principal so and so refused to talk to us about this very difficult issue'. And they will get someone else who will talk but is probably not as knowledgeable as you are. Most of the time, and legitimately so, they will reach for other voices besides yours also. It is fun at the time because the reporters you see on TV, they come along and they talk to you and they are articulate and highly

capable and well informed. They will dig into things and you will respond. Excited they go out to their truck to edit a half-hour or 20 minute interview into a 10 second presentation. They entreat you to watch at five or six o'clock, which you do. They present the story that they set out to tell in the first place. Not your story — but their story. This is a highly sophisticated process and it has power and impact.

Another means for informing the wider community is by permitting scholars and researchers to involve the district in the conduct of research studies. This activity was controlled through the research and evaluation unit. Mr Berry describes one of his experiences with this activity.

We were struggling with desegregation in the Riverside Unified School district. Fritz Mondale was heading a sub committee which was rewarding people for obeying the rules and penalizing those who volunteered, and paying those who were waiting to be forced. I let Mondale know. He rewrote a section of the initiative and put 14 per cent of the money into reward for volunteering. During the time that this law was being implemented, a huge study was being conducted here. This school district study initiated by professors on the local campus collected more personal student data than any other study prior to that time and most other studies since that time in any public school research. The media, anxious for this data, began to report. Shortly thereafter, I got a call from Senator Mondale and a Los Angeles representative of the Associated Press who claimed there was a report out on the East Coast that quoted me that integration does not work. Mondale asked, 'Ray, what the hell is going on out there?' In tracing back where this information could have originated we discovered a Harvard professor reached these conclusions from a request for funds that had been sent to Washington. In our request, we had listed problems which this guy, who seems to be quite unprincipled, took and because it had our name on it used it under the guise of a research report and claimed failure. After clearing this up with the Senator he took it to Harvard and the professor got fired. Incidentally, we never received a retraction of this report.

As can be seen from the example presented above, once faulty information is out in the public, it is virtually impossible to retrieve or revise it substantially. Thus, the question of the vulnerability of school districts and superintendents turns out to focus on how information travels. What is the process, whose responsibility is it to provide the media with sound solid data, and what are the processes for communicating with the rest of the public effectively?

Generalizations

This chapter has focused on how the control of information served to avoid conflict in a school district. The conflict was averted because the superintendent controlled information through three check points directly under him. The first was the research and evaluation unit, the second and third were the staff and intellectual's position in his office.

Mintzberg (1979) claims that the decision process is most centralized when all of the steps are controlled at one point or with one person. These steps are: collecting information; analyzing it; making a choice; needing to seek no authorization of it; and executing it.

A review of these steps in regards to Mr Berry is instructive. Mr Berry collected his own information through various means, but the most obvious was the institutionalization of the research and evaluation unit which was directly under him. The director presented the information and initial analysis. Out of that analysis and presentation, Mr Berry chose what to report, when, to whom, how, and who would report. He sought no authorization for this presentation. Ultimately, he was able to execute the information himself. Information not processed through the research and evaluation unit was processed in Mr Berry's office. He had a staff member who reviewed the local press' news clippings on education, for example. Information which was new, novel or for some reason not appropriate for being processed through the research and evaluation unit was handled by the intellectual, reported in Chapter 3. This intellectual prepared official, technical and theoretical reports for Mr Berry. He also prepared graphs, illustrations and other support materials for the various groups as the need arose. Importantly, he also linked the academic and research institutions to the school district. In sum, the control of information was most assuredly under the superintendent.

How does this control of information reconcile with the claim that the Riverside Unified School District had changed from a bureaucratic to a decentralized organization? The explanation lies in the broad application of Levy's (1966) definitions of decentralization and centralization. Organizational units are decentralized to the extent that they make decisions regarding their particular areas. To reiterate,

'The major structure of decentralization in the *last analysis* (our emphasis) is always the actual if not ideal autonomy of the nearly self-sufficient units within the sphere of their self-sufficiency' (p. 100). The organizational issue which arises is how to *protect* this structure. Mr Berry chose to protect it through the control of information. Teachers were permitted autonomy, principals were likewise permitted autonomy, as were the other groups. Conversely, they were also inhibited from infringing on each other's areas. For example, principals could not dictate to teachers in the same way they had under the bureaucracy. This case, thus, serves as an example in which decentralization succeeds because the superintendent retains control of the total organization through the control of information. Most importantly, decentralized autonomy does not mean losing sight of the organizational mission. It means that each unit is allowed to contribute to the mission in its special way. Mr Berry intended to improve children's achievement and school staff's work performance. Information provided the basis and rationale for the entire organization for the execution of this intention. The reason the control of information is critical for the leader is as Glasman (1986) and Kanter (1983) explained, information is the critical tool in the decision making process.

Schlechty and Joslin (1986) identified two elements that should never be decentralized and one that cannot be decentralized. The establishment and articulation of superordinate goals and binding myths, articulation of values and commitments, and the reinforcement of these values and commitments in behavior as well as words flow from the top. What the school is about, where the school system is going, and what problem originates, emanates, and resides in the superintendent's office.

A second responsibility that cannot be delegated is responsibility for bottom-line results. The quality of performance of the work force in schools is the responsibility of the chief executive officer. Test scores and other measures of student achievement are like profits in a large corporation. As in corporations, where the chief executive officers, rather than the first line supervisors, are held accountable so should they be in schools. What the first line supervisor is accountable for is doing those things that top management believes will produce profits. If first line supervisors do those things well, they are rewarded. If those things do not lead to profit and growth in the long run, it is the CEO that is fired, not the first line supervisor (see Schlechty and Joslin, 1986, p. 159).

Mr Berry, the superintendent of the Riverside Schools, understood this aspect of his organization very well. His contract with the school board was a yearly one, which served to enforce the accountability. The structure he created and the control of information were designed to benefit teachers' performance. This insured that the one element that cannot be centralized regardless of strenuous effort to do so would be protected. 'Problem solving [instructing children] is best left to those whose hands-on-experience and expertise provide them with the advanced knowledge to invent novel solutions' (Schlechty and Joslin, 1986, p. 159).

The dilemma of structuring a bureaucratic organization into a decentralized one with vast control of information was solved because another important aspect was attended to. Mintzberg (1979, p. 289) claims that the 'two most effective means to control an organization from the outside are (1) to hold its most powerful decision maker — namely its chief executive officer — responsible for its actions, and (2) to impose clearly defined standards on it'. Mr Berry intentionally assumed responsibility for children's improved achievement and staff performance and his standards were clearly defined through the decentralized units. There was no need for this organization to be controlled from the outside. Mr Berry controlled it, primarily through the utilization of a sophisticated information system.

There is an important organizational consequence of the structure just presented in this chapter. Decentralized units directly under the control of the superintendent developed a loyalty and obligation to the superintendent and organization rather than to their immediate group or other groups. The most obvious example is the principals. Even though some of the principals opposed Mr Berry's directions, they could never marshall enough influence or power to have an effect in disturbing the organization. The control of information through decentralized units by the chief executive appears to have both practical and theoretical potency.

Chapter 6

The Relationship Between Intentional Leadership, Change Processes and the Nature of Decentralization

Leadership Studies

The literature on organizational leadership has focused on personal attributes and organizational contingencies. Currently, the literature has focused on constructs based on culture. The research has reported that leaders normally begin with a vision and sense of mission which, then, they are able to transmit throughout the organizational structure. Most research has been unable to link leaders' visions and the effectiveness and efficiency of organizations through organizational acts. The construct of culture has been useful to describe the nature of this linkage. Many descriptors and metaphors have been tied to successful organizations and leaders, but the research reports have equivocated on the specific actions leaders perform. The organizational development literature has created programs to facilitate leadership effectiveness, but again, the concern has been lodged primarily on the *programs* rather than on the *actions of the leaders*.

In regard to the literature on the superintendency, the findings show superintendents' vulnerability, and conflictual and political context as the critical factors, rather than the superintendents' actions to improve the organizations', students' achievement, and staffs' work performance. A brief synopsis of this literature is presented in order to provide a comparison to the present study which demonstrates the linkage between a leader's attributes and acts and organizational effectiveness.

The literature of the 1960s and 1970 on school decentralization, shows that school organizations' reform efforts were, for the most part, not initiated by the superintendent. Rather, they were imposed by external parties: the community, or interest groups within the community. Thus, the initial demand for organizational effectiveness did not originate with the superintendents (see Mohr, *et al.*, 1976; La Noue and Smith, 1973; Ornstein, 1974; Levin, 1971; Fantini and Gittell, 1973; Ziegler, *et al.*, 1973; Lawson, 1973; Steinberg, 1974; Ravitch and Grant, 1975; Steinberg, 1977; Bard, 1972; Borman and Spring, 1984; Wissler and Ortiz, 1986). Not surprisingly, this research indicates that external groups were granted access to participate but not to make decisions, thus, their effect on the organization was minimal.

A few decentralization studies demonstrate how the superintendents initiated school reform. Gittell (1967a and b), Rogers (1968), Sizemore (1981) and Wissler and Ortiz (1986) and others report how central offices could remain unchanged if new positions such as area or regional administrative posts were available to persons who could *represent* the school district but *not speak for it*. Hughes and Bratton (1978), Cawelti (1974) and Patterson and Hansen (1975) likewise reported how these newly created posts were bypassed when decision issues occurred.

One study (Wissler, 1984) shows how school site budgetary autonomy could be granted to principals in a successful decentralization effort. Patterson and Hansen's (1975) survey research reports how 'building level personnel [principals] had the *most influence* on virtually every issue' (p. 27). Chapman and Boyd (1986) report how a school district in Australia is attempting school site decentralization. This extent of decentralization, however, appears to be rare. Superintendents are not reported as granting autonomy to school sites or to other school units. Rather, they are reported to have retained the essential decision making apparatus for themselves and the central office staff.

The superintendents' reluctance to decentralize is explained in several ways. First, most decentralization efforts were analyzed through the imposition of Allison's (1971) decision making models with their attendant assumptions. The political bargaining, pluralistic bargaining model assumed the process was a political one, with the superintendents' political skills at stake. The organizational process and rational man models assumed the process to be a rational one with the organization's and/or superintendents' rationality to be tested. Peterson's (1976) ideological model assumed the process to be based on ideology. When all of these bases were found to be absent or all present in the same decentralization effort, researchers were unable to explain why decentralization failed to take place.

The case of Sizemore (1980; also reported by Arnez, 1981) differs in that even though the decentralization process was initiated by external parties and Sizemore was hired specifically to decentralize the organization, each of her decentralization proposals were rejected whilst the individual or group board members' proposals were temporarily implemented. The structure which was created is instructive to review (see Wissler and Ortiz, 1986). An administrative unit was created specifically to link the community with the school district and decentralization began through this mechanism. Upon its abandonment through budget cuts, the process of decentralization moved toward administrative decentralization. Several attempts are reported: a six unit regionalization was created; this was subsequently increased to twelve units; administrative authority was delegated to these decentralized regions during the third phase. Finally, delegation of policy making authority to the decentralized regions took place. The consequence of all of these actions was conflict. The school board and Sizemore, the superintendent, struggled for control. The analysis of this process (Sizemore, 1981; Arnez, 1981) lays the burden of the failure to decentralize the Washington, DC schools to the inability of the District of Columbia Board of Education to formulate a conceptualization of decentralization or to chart a strategy for the school system to adopt. This study graphically demonstrates the influence of the political and conflict-laden context upon the efforts of superintendents to decentralize.

An example of a study which fails to report how superintendents grapple with the issue of reconciling decentralized units within a school district is the Rogers and Chung (1983) assessment of the New York City school system's 1970 decentralization. They report eight school districts in which decentralization was effective. They found that the students' reading performance improved, and that effectiveness was in large measure determined by the presence of a 'strong' superintendent. Strong superintendents *emerged* where educational politics were stable, where a strong leader could insulate the superintendent, where there was an established parent and community organization, and where a coherent, long-term approach to educational programming and staff development could take place. We are left, however, with the question, 'How does the New York City School System's *superintendent* manage the decentralized and non-decentralized units in order to provide educational services throughout?' unanswered. The decentralization literature, thus portrays superintendents reacting to community pressures and school board demands. With the exception of the Wissler (1984) study, these superintendents work in highly political and conflict-laden environments where success is difficult to achieve.

The recent studies portray superintendents as *victims* or *vulnerable* to their organizations because of conditions beyond their control. For example, Borman and Spring (1984) claim that because schools receive funding on the basis of meeting externally determined regulations at the board, state and federal levels, school leaders learn to gain these resources primarily through conformity to regulations. They write, 'The orientation of the members of the bureaucratic structure of schools is not toward responding to the needs and desires of the public or market place but toward meeting the requirements of external regulations' (p. 88). Rowan (1981, p. 49) writes, 'As administrators turn increasingly to the demands of conformity, they devote less time and energy to core instructional problems'.

A recent newspaper article (Willis, 1987) quotes the local superintendents' descriptions of their work. The article begins with the recruitment announcement: 'Wanted: top educator to be school district superintendent. Experienced in office and community politics. Able to communicate to board, employees and parents. Able to solve problems and understand finances. For job conducted in public view' (p. B-1). In the same article superintendents being pressured to resign are advised to 'cling to your contract' and 'go for the gold' in negotiating a contract buy-out with the board. Several superintendents from the surrounding school districts' resignations are listed to demonstrate the precariousness of the job. Following that, superintendents' descriptions of the superintendency are presented.

Out of twenty-four descriptors provided by the superintendents, seven relate to the precariousness, insecurity and conflict within the post. For example, one superintendent says that when a person decides to become a superintendent, he or she also should accept the job's *insecurity*. 'You have to realize you are one of the most visible public figures in the community, and that makes you vulnerable to those who disagree with you. Superintendents don't make policy, they administer it, but many people cannot see the difference. The superintendent becomes the whipping boy for society's problems'.

Five of the descriptors relate to the importance of trust, harmony and cooperation between the superintendent and the various parties. One superintendent says, 'It often boils down to a matter of *trust*. When the trust relationship between superintendent and board evaporates, both parties are better off if they make other arrangements'. Another claims, 'It is the responsibility of both the superintendent and the board to see that they get along. You need a *match* between the board and superintendent, and both are responsible for seeing a match occurs. If it does, *harmony* is produced and that eliminates the precariousness of the situation'.

Eight of the descriptors relate to the challenges of the job. For example, one states, 'I like having the ability to conceive a plan, a vision, a direction or a dream, and to develop and implement it, and see it succeed'. A superintendent who has survived several board recalls says, 'You have to do your job as you see it. I survived because basically I'm a direct and open communicator. I don't have any secrets. If the answer has to be "no", say "no"'.

The other descriptors cover issues related to orientation and expectations, and team work. For example, one explains, 'In any organization, they look at the chief executive officer and have *expectations* of performance at a very high level, and I think that's appropriate'. These descriptors reveal the continuing concerns of these present day officials with an edge towards the positive. Some superintendents view their work more as challenging than precarious. Researchers, however, have tended to emphasize the conflict and vulnerability of superintendents rather than their resourcefulness and initiative (see Cuban, 1985). For example, Blumberg (1985) writes that 'by 1890 a variety of forces had been set in motion that were to affect the character of the superintendency. Primarily what was to occur was a shift in the relationships between school board and superintendent, particularly with regard to issues of executive power and management prerogatives' (p. 26).

Cultural Considerations

Joseph Mayer Rice wrote in 1893:

> The office of the superintendent is, in my opinion, one the importance of which cannot be overestimated. Indeed, in the study of the educational conditions in any given locality, the superintendent may be regarded as the central figure — a careful consideration of what he is, what he does, as well as the circumstances under which he labors, will scarcely fail to point out the reasons why the schools of that locality are on a comparatively high or low level. When he is a thorough educator — that, when he has made a profound study of the science — spares no pains in instructing his teachers in educational methods and principles, and if fully sustained in his

actions by the board of education, the schools in his charge, if there be not too many, improve rapidly and ever continue to advance. But a modification of any one of these conditions impedes the progress of the schools ...

But when the superintendent labors under very unfavorable conditions, he seldom fails to stamp the schools with his individual pedagogical ideas, thus giving the education in his schools at least a tendency in a certain direction, provided·he remains long enough — say four or five years — in any one city (pp. 11–13).

Hoyle and McMahon (1986), in their analysis of the superintendency in cultural institutions, present a current profile:

As institutions of cultural transmission the management of schools is, or ought to be, of a different order. It is concerned with a much more diffuse form of leadership related to the creation of meaning rather than to profit and loss. The symbolic function of school leadership may be met by a leader of vision and charisma, but it would generally be held that vision is preferably informed by a good group of the realities of the educational world: forces in the situation, subordinates, the environment, as well as forces in the leader himself or herself (p. 22).

A small body of research, primarily that concerned with public participation in school affairs (see Lutz and Iannaccone, 1978; Ziegler, *et al.*, 1973) has linked superintendent effectiveness to the value orientations of the communities and school districts in which they hold office. The history of the Riverside Unified School District is a good example to show how the superintendents retained long tenures because they reflected the predominant values of their period. A short synopsis is presented to illustrate this.

Superintendent Success and Community Values Congruency

Education has been important to Riverside since its founding. A school was opened in one room of a small pink house on the corner of Sixth and Lime in 1871 by a widow named Mrs Rogers. This was less than a year after the colony was founded (Greater Riverside Chamber of Commerce, 1983, p. 19). Ten pupils were enrolled. By 1873 forty of the Colony's 300 children were in school. In 1875 the second school was built on Central Avenue (Sunnyside School), and more rooms were added to the Sixth Street School in 1876. In 1881 construction began on what was to become Magnolia Elementary School, and in 1888 a three-story brick school opened with the third floor serving as the high school; the same year Highgrove Elementary School was built. In 1890 the high school graduated its first class of seven students (Report Card, I, 1977, p. 2). The district was formed by the forward thinking of a Riverside school trustee, E.W. Holmes, who with others 'lobbied for a high school district, and legislation was passed in 1893 which permitted state funding for high school buildings and programs' (Covel, 1977, p. 14).

The period from 1870 to 1900, or the pre-superintendent era, was a time of lay leadership for the schools in Riverside. Public education was managed by leading citizens as a public trust. These trustees (school board) interviewed and hired teachers, and 'policies, plans, and buildings were promoted by the community residents through the trustees' (Covel, 1977, p. 38). Teachers who were called principal-teachers were placed in charge of one-room schools or of a school with several teachers. The formalization of the principal role in Riverside did not take place until 1899–1900, and from there it spread to the rest of the county. Principals were teachers who were released from teaching duties and were the forerunners of superintendents (Covel, 1977). Under this arrangement lay trustees retained the decision making right for the schools.

Four Superintendent Profiles

From its beginnings and through the time of our study Riverside had four superintendents. Arthur N. Wheelock in 1902 was first. The responsibility for nine elementary schools and one high school began the move from lay trustee to professional school control. The centralization process which was to increase through three superintendents until 1960 began at this time. For example, in 1909 the Riverside Board of Education decided that 'filling of vacancies in the teacher force be left to Superintendent Wheelock, with power to act' (Covel, 1977, p. 43). 'In 1911, he assumed recruiting power for the high school principalship and the following year he gained the power to fill all vacancies' (Covel, 1977, p. 120).

Eight schools were contracted during his superintendency, including three junior highs and a manual training building. Mr

Wheelock was an advocate of manual training and of the separation of the sexes. From a copper-box time-capsule bulldozed up in 1965 from Polytechnic High School came a copy of his October 1911 plea to the board:

> Granted that there is something to be gained for both boys and girls from working together. But ... our schools ought not to produce girlish boy or boyish girls. Something of virility is lost to the boy whose high school life is passed in a school in which girls and female teachers predominate. It is even a more serious matter for the girls. Something of the bloom, the charm of girlhood and young womanhood is endangered in the rather boisterous atmosphere of a mixed school (*Bulletin Board*, November 1965, Vol. 12, p. 2).

Patterson (1971) said Wheelock was 'a particularly able administrator (and) ... he was thoroughly representative of the educational trends of the time, both in respect to its virtues and as to what would be eventually described as its narrowness'.

Wheelock's strong and long superintendency of 26 years left Riverside with at least two legacies. One plainly visible to all was the Wheelock Stadium named in his honor. The other, Patterson calls the 'undeclared Wheelock policy of placing schools and drawing boundaries with the direct effect of increasing segregation' (1971, p. 481).

Wheelock's superintendency began and/or increased four types of exclusion practiced by the school district. They were:

1   Lay citizens no longer took the active role they once had in choosing teachers and principals (Covel, 1977).
2   Females were excluded from secondary level principalships from 1912 until 1974. (In 1912 under Wheelock, 'Eugenie Fuller was voted out of her job as high school principal' (Covel, 1977, p. 120).
3   Male and female students were directed into some and excluded from some other classes, such as manual arts (Covel, 1977).
4   School boundaries were drawn for purposes of exclusion on the grounds of race (Patterson, 1971).

Since Wheelock's 'direct influence extended for 40 years', it went well through the term of Mr Ira Landis, Riverside's second superintendent.

Landis, an insider who was promoted from an elementary principalship in the district, continued the tradition of long tenure by serving 23 years. One person who knew him stated:

He was a smiley-faced man but he hadn't the vaguest idea about education *per se*, but he sure did know about people. Because he felt inadequate he hired an assistant by the name of Carl Cress. He knew all the instructional things; it was really very smart of Mr Landis to hire Dr Cress.

The district hired Dr Cress as curriculum director in 1944 and moved him to the post of assistant superintendent shortly thereafter. He served 17 years under Mr Landis in that position and held the first doctorate in education in the system (Covel, 1977, p. 81). The data show that the hiring of a man with a doctorate was a two-edged sword for Superintendent Landis. A principal reported that:

As Dr Cress's ripple effect was felt with him shoring up the instructional end, I think it made Mr Landis quite nervous and nervousness brings on dicator-type things. '*All right, you will. Everyone will!*' Well, I think that's what Mr Landis kind of did. Mr Landis was old-time but very structured because I think he got nervous seeing all the new things coming in.

Landis served through the years of World War II, and under his direction one school for the handicapped was established. With the end of his 23-year term in 1951 an era of agricultural life was also ending. Though Landis served through a time when only a small amount of growth occurred in the addition of new schools, he, too, left his mark on the system.

With his hiring of Dr Cress in the district office, centralization was further accomplished. Cress's having a doctorate and his 'shoring up the instructional end' meant that the central office gained greater influence over instruction and curriculum. Coupled with increased central office strength was Mr Landis' personal style which was described as 'structured' and which at times even pushed him to 'dictator-type things'.

Bruce Miller, the third superintendent (1951–1968) came to Riverside with a masters degree from the Claremont Colleges and eleven years' experience as the superintendent of schools in Ontario. As Riverside's superintendent he presided over the explosive growth which added 19 new schools (14 elementary, 2 middle, and 3 high schools) to the system. When Mr Miller became superintendent, the high school and elementary districts were separate, and Riverside Junior College was also under his supervision. The budget was $3.5 million and there were 486 teachers (*Riverside Press Enterprise*, September 15 1982). During his term in the 1950s and the 1960s Riverside nearly tripled its student population to its peak growth of 28,000 students (Report Card I, 1977, p. 4). The budget during his tenure in office had grown from $3.5 million to $18.5 million.

During his term of office until 1965 he, with the board, continued the unofficial Wheelock policy of 'adjusting boundaries so that large groups of minorities were excluded effectively from the white schools' (Hendrick, 1968, p. 33). Minority persons interviewed said, 'We didn't have that much rapport with him'. For example, he was advised (possibly by the police) not to attend a meeting on Friday, September 10 1965, in which the President of the Board was to speak to the minority community. Hendrick reported that 'Miller remained at home with hurt pride' (Hendrick, 1968, p. 107).

However, it was during his tenure (1965) that Riverside City Schools 'became the first school system in a city exceeding 100,000 in population, and with a total kindergarten through grade 12 enrollment of more than 20,000, to develop and implement a full-scale racial balanced plan' (Hendrick, 1968, p. 1). The plan called for the closing of three schools and the extensive transportation of minority children into nearly all-Anglo schools. Miller saw desegregation through and served three more years.

Miller was described 'as one of the most extroverted personalities I ever worked with'. People who knew and worked with him described him as 'hail fellow well met, a jolly back-slapper who loved to play cards'. Another said, 'Mr Miller was a very gregarious, personable kind of man, outgoing, and he loved attention'.

As an organizational leader he continued and tightened the centralization which had begun under his two predecessors. Common descriptors of the system under Miller were 'traditional', with a 'board supporting administrators who basically ruled the organization from the top-down'. It was a typical one, yet rigid, autocratic, and almost militaristic. They said the statement was a little too glib but it was 'top-down'. Some said that 'edicts' and 'mandates' were passed down with ease so that 'we could all be alike'. Others said the district was sophisticated, meaning that the district was full of strong professionals in the central office who knew and performed their jobs. They had worked up through the ranks and produced a program no one could really quarrel with.

The Riverside Unified School District developed under Miller as an academic leader throughout the state. For example, an administrator reported how it pioneered the program of driver education and later learned that the state legislature had recognized the need for the program and was funding it by incentive.

Miller's love of attention and desire to 'enhance the district' led him to work at the state and national levels. Though he was a traditionalist in structuring the school district as a bureaucracy, he had 'a deep sensitivity and awareness of people'.

To sum, Miller was a gregarious, organizational autocrat. He ran a classical, hierarchical, militaristic pyramid of an institution which was successful in that it was known throughout the state as a fine academic school system. Following desegregation the district became known throughout the world for its 'almost spectacular' resolution of the issue (Gerard and Miller, 1975, p. 26). And at the same time 'as the symbol of an intensely distrusted school authority, he [Miller] represented to some much of what was wrong with the schools' (Hendrick, 1968, p. 107).

When he retired in 1968, the Board President told Miller he had 'headed the district through more changes and more problems than your predecessors in 40 years have had to face' (*Riverside Press Enterprise*, September 15, 1982). After thirty-eight years as a school administrator, seventeen in Riverside, he retired a happy man. His last years had been perhaps his most challenging but probably the most rewarding and distinguished for him (Hendrick, 1968). With his retirement the district had had only three superintendents in sixty-six years.

The fourth was Mr E Ray Berry who had left a superintendency to come to Riverside as personnel director in 1960. In 1961 he became an assistant superintendent to Mr Miller in the areas of instruction and personnel, and in 1962 he became the associate superintendent. A graduate of UCLA, he continued his ties with that university while he was with the Riverside District. In 1968 when Mr Miller retired, the Board President described what happened:

> When Ray was selected as the superintendent, the typical fashion would have been to get a committee and to get consultants to bring in candidates and so forth. Ray had been our assistant for three or four years and everyone thought we should just ask Ray, and so I got him on the phone and it was just like that. So we had a lot of faith in him.

A direct appointment reflected the faith the board did have in him. When the board selected Berry, they were hiring a superintendent who was in some ways the direct opposite of Miller in personality and personal style, as well as an organizational leader. As a person he is described as a 'low-key personality' (*Riverside Press-Enterprise*, March 30 1978), 'able to listen with the trust and admiration of people'.

Irving Hendrick tells a story which demonstrates how deep the trust of him was in the community. On September 12 1965, Riverside was facing a boycott of its schools by the minority community. Wilson Riles, the future State Superintendent of Instruction, became involved:

> He received a call from a Negro lady visiting in Riverside. In addition to this conversation, and his conversations with Miller that same day, Riles talked by telephone with one of the boycott leaders who explained in greater detail what was occurring. It was clear to Riles from these conversations that the Negro community completely mistrusted the sincerity of school officials. The school district intended to do nothing, the boycott leader told Riles. In the course of that conversation Riles asked pointedly if there was anyone in the administration you trust. 'Ray Berry', was the reply (Hendrick, 1968, p. 110).

Berry's personal style was not only different from that of Miller, he also faced different problems. The decade of the 1970s saw the city grow in population, but the school district began declining in enrollment. One high school was opened in 1973 during his tenure, and the district administration sold and moved its offices from 3954 Twelfth Street to Fourteenth Street. The move was made because less growth took place than had been planned. 'The sale resulted in nearly $1,000,000 in savings to the district. Use of these funds permitted the Board of Education to update and improve the district's buildings' (Report Card 4, 1977, p. 1).

His tenure was a time of controversy over textbooks, sex education, drugs, discipline and dress codes. Larger educational issues such as how to educate effectively each child to his/her highest potential and serious drop-out rates were concerns in Riverside too. It had desegregated its schools and then faced the problem of integrating its classrooms.

Berry differed sharply from Miller in some ways in his organizational style of leadership. As an administrator, people reported 'it was not his style to mandate — he wanted to initiate things, to open doors for people to create, to let things evolve'. Berry said that he could not see how an organization could be run from the top down without almost 'guaranteeing mediocrity'.

The two superintendents, Miller and Berry, were described by one person as, '. . . quite a team. They were very different. Both were very personable. Everybody liked each one of them'. They differed in age, experience, personality and philosophy on how an effective school district should be governed. Berry inherited a structured bureaucratic district which was highly centralized, but he did not leave it that way.

One school official said of Berry, 'He was out to break a stifling framework and he ripped it wide apart and he got people creating'. Berry was not a centralizer. Under his superintendency the centralizing trend of his three predecessors was reversed.

The three superintendents who served the Riverside Unified School District were individuals who reflected the pervasive values of their time. When Mr Berry came on board, his values were to change the organization from its bureaucratic character to a decentralized one. This was acceptable to the school district and the community.

The Differences Between this Superintendent's Leadership and Others

Mr Berry differed in his leadership in several important aspects. First, Mr Berry assumed his position with full support of the board. He was known by the school district and the school board members and he had helped the previous superintendent desegregate the school district successfully (see Ortiz and Hendrick, 1987). This broad support was important in his ability to *internally* initiate the changes towards the improvement of his organization.

Second, Mr Berry began the district's changes with a broad, uncontestable purpose. He committed himself to improving the academic achievement of students and the work performance of the staff. Each of the changes he made were linked to one or both of these objectives.

Third, he did not claim to be *decentralizing* the school district, or changing its structure until it was so. Aside from Miller's recorded (Riverside Unified School District, 1960–1979, 2–27–67) use of the term in recommending program adjustment and in connection with saving money, when the *principle* of decentralization was accepted by the board, there are no other written materials, such as proposals, programs, or intentions regarding decentralization. This is an important consideration when this decentralization style is compared to Sizemore's (Sizemore, 1981). Sizemore prepared decentralization strategies which were subsequently attacked and rejected. Nothing was written in the Riverside Unified School District until 1971.

Fourth, the decentralization of the Riverside Unified School District was primarily directed at *improving* educational services rather than *extending* power to groups heretofore excluded. The process of inclusion in the Riverside School District was to deliver educational services to children rather than for personal or group benefit. This orientation was so pervasive that Mr Berry accepted only a year-by-year contract with the Board, fully accountable for his accomplishments.

Fifth, the profile of the school district was enhanced, highlighted, and displayed, whereas Mr Berry, the superintendent, remained a person behind the scenes. His intention was to accentuate the organization rather than its leaders.

Sixth, Mr Berry was able to decentralize the school district without damaging conflict, because he used *information* as the tool in making changes, rather than persons. He manipulated, controlled and sought information rather than people. Because the nodules of information were directly under him, each decentralized unit could be protected as it moved from a bureaucratic context to a decentralized one.

Seventh, the quality of the participation of the school staff was the primary concern of this superintendent rather than programs. Also, educating the children was his focus rather than implementing programs. These objectives necessitated decentralized units which could solve their problems as they came up, rather than centralized units which needed to be supervised.

This leader, thus, as an intentional one, did not dwell on the political vagaries, or social turmoil. Instead, he acted to permit his school staff to deliver educational services to the best of their abilities. Because the intent was clear and his style was non-fluid (Ramos, 1981) he could make dramatic changes with the full support of the school board and the community. He was trusted because the benefits to be derived were for the organization and not for himself or favored ones.

Some of the indicants of the quality of participation are the metaphors which became common during this period. The bureaucratic metaphor is military. A military centralized district has a chief officer in charge who commands a hierarchy of graded levels. Everyone in these graded levels has a rank. Those higher are superior officers and either line or staff officers. Those persons of lower and lowest rank are subordinates. There is a chain of command through which orders travel. Subordinates are issued orders and commands which are to be obeyed. Moving outside the chain of command or failing to follow an order is insubordination which may result in reprimand or dismissal.

A member of the organization fights to get to the top and then he strives to win out over his competitors by the use of tactics in the form of trade secrets. He maneuvers to outflank the competition. He targets the competition, zeroes in on the market, and pins it down by a barrage of advertising. In these few sentences we have corroborated Weick (1979) in the use of twenty-three military organizational descriptors.

During the period of Berry's superintendency, the military metaphor was replaced by one of flight. The data show only two competitive, and one military word. A teacher said, 'We started trying innovative things to make our school better than other schools'. And a principal said, 'We competed to have our projects funded'. Instead of a leader who issued orders, Riverside had one who challenged principals and staff to 'reach for it'. The organization was not seen as ranks of levels but was 'opened up' to the participation of all. There was no second class section for a subordinate group. The leader talked about drawing 'top flight people into education and releasing their power'. A principal quotes him as saying over and over, 'We have top flight teachers here'. In a class he was teaching he stated that when you find 'top flight minorities and women hire them, but you can't keep them because you will find yourself conducting a farm league'.

Instead of a leader who increased the number in his high command, he decreased the central office staff and endeavored to teach them to become a 'service not a command structure'. When he spoke of evaluation he said that the weak administrator is not the one 'who will take off' and leave at the first hint that his performance is lacking. He continued that as a superintendent 'you reach for good people who will open up the system, express themselves and start reaching for the stars'. He thought that the right kind of system which called for that kind of leadership pulled people into it and 'that people who found security in bureaucracy found a way to get out'.

Once people were in the organization they were not constrained by tight rules and regulations but they were told by Berry to 'fly your wings', and they were expected and allowed to make mistakes without reprimand. He often talked about 'building a program and reaching for restructuring a better learning program, but it has to come out of the people, doesn't it? You can't say, "you haven't done a good job"'. In his role of the superintendent he 'helped move these up' or 'moved the whole thing up'. Sometimes that was done by conducting 'interventions — that is, a massive process of getting in the way of what was going on'. Communication was to be 'opened up'. The Riverside cluster system was developed so 'you could see these people in a different context and communication could be opened up'. He described the result of the cluster organization (which had no levels or ranks) as the 'programs took off'.

Instead of commanding a lockstep program the school district had encouraged '30 to 40 satellites flying in every direction'. And instead of looking for persons who were expert at taking orders, Berry advocated 'powerful decision-makers because it was like tying your tail to a kite and you've got to be prepared to go with it!' One principal stated that with the autonomy and freedom in the district you had the feeling that 'the sky is the limit'. And lastly, Berry said the purpose of the district was 'to create blue sky experiences for children'. These metaphors provided a context in which the school staff and students perceived the schooling experience as a challenge rather than as a political, contested, or mindless activity.

Organizational Change

There is a substantial body of research on organizational change. Most of this research indicates high levels of conflict, ambiguity and failure. Several aspects are associated with these results. One of them being that when organizations change, so do the roles of the school person- nel and that organizations, or the administrators of the organizations fail to provide training and opportunities for persons to change their roles.

Fullan (1982) explains how educational change takes place. He writes:

> The district administration mainly determines whether dis- trict-wide change gets implemented. Whether central admin- istrators are equal to the task is sometimes beyond their control but less often than many administrators have shown. Being equal to the task, as stated above, means integrating three things: addressing *technical knowledge* requirements; pos- sessing a *conceptual and technical understanding of the dynamics of change*, which guides and generates one's actions and reactions; and having the *interpersonal skills* and behavior of an active communicator, who gets around and demonstrates the sincer- ity of one's intentions as well as knowledge of the problems of change faced by system members (p. 168).

In the case of the Riverside Unified School district the superintendent determined that the total school district would be decentralized. The integration issue was facilitated through the decentralization process and the control of information.

In the case of school reform, the issues have generally been that teachers have not been able to change from their ability to teach middle class children to their ability to teach diversified student groups. The most pervasive focus, however, has been that different children require different *programs.* This concern has necessitated school administrators, including superintendents to direct their efforts on programs rather than children. Many of the school reform efforts in the 1960s and 1970s were concentrated on *implementing programs* rather than on *teaching children.*

The present study indicates very little mention of programmatic interest. The organizational changes which took place were those which *released* teachers to *teach,* and *facilitated* school personnel to *teach* and *deal* with *children.* The conflict which arose during this period is also minimal. Two fundamental characteristics stand out in this case of dramatic organizational change. The first is the length of time it took to decentralize the organization. A total of 18 years was required to decentralize a medium-sized urban school district.

The second is the quality of participation of the organizational members. Mr Berry believed that every school staff member's responsibility was to deliver educational services. He also believed it was his responsibility to facilitate that process. Every member, was thus given a purpose in the fulfillment of the organization's mission. Teachers' roles were enhanced and the proof was that they could participate in the same group as the principals and central office staff. Teachers' concerns were given primacy and all organizational members' efforts were directed toward a common goal. The process by which the change took place was as important as the change that was occurring. Decentralization took eighteen years to accomplish because Mr Berry systematically instructed the organizational members as the changes were occurring. The changes were incremental and gradual rather than traumatic or sudden.

Another aspect to this quality of participation is that most of the organizational members felt a sense of challenge, excitement and engagement as the changes took place. This is particularly remarkable when it is realized these changes took eighteen years. Sustaining interest and momentum for this length of time is rare in organizational change. Part of the reason is that the changes did not lose sight of their purposes. The changes were necessary in order to improve children's academic achievement and school staff's work performance.

These goals are organizational rather than personal. The organization was, thus, able to sustain enthusiasm and purpose. Lawler (1986) writes:

> employee involvement in decisions that affect their work situations can be an effective way for devising better work methods and for solving important problems. For this to be truly effective, however, three conditions need to exist. First, employees need to be knowledgeable about the issues that they are devising solutions to. Second, the employee must be motivated to solve the problems in a way that is consistent with the best interests of the organization. And, third, mechanisms must be set up to facilitate the implementation of the solutions (p. 35).

The same researcher and others (Baldridge and Deal, 1983) advocate that 'the potential effectiveness of a participative program is dependent on how much it moves rewards, power, communication, and knowledge downward. Associated with this downward movement is simultaneity'.

Consistent with what is reported in the successful decentralization of the Riverside Unified School District, Lawler claims that 'organization-wide change in a large organization ... can take decades ... Overall moving to participative management is a long, slow, somewhat chaotic process' (p. 233). It is the participative management part that determines the quality of participation of those engaged in changes.

Schein (1985) presents five mechanisms by which founders or leaders are able to embed their own assumptions in the on-going daily life of their organizations: '(1) through what they pay attention to and reward, (2) through the role modeling they do, (3) through the manner in which they deal with critical incidents, (4) through the criteria they use for recruitment, selection, promotion and excommunication they communicate both explicitly and implicitly the assumptions they really hold, and (5) through the conflicts and inconsistencies which get communicated and become a part of the culture' (p. 242). He concludes, 'Leaders do not have a choice about whether to communicate. They have a choice only about how much to manage what they communicate' (p. 243).

Kanter (1983) found that innovation (change) is inhibited by segmentation in which the organization is finely divided into departments and levels, each with a tall fence around it. Information is a secret rather than circulating commodity. In explaining the difference between centralized and decentralized organizations regarding change she writes, 'There seems to be more initiation of innovation in decentralized systems, but much less adoption' (p. 405).

The Nature of Decentralization

As stated earlier, decentralized structure according to Levy (1966) means self-sufficient units' autonomy within their own sphere of self-sufficiency. For schools, this is a critical point. One ever pervasive criticism of educators is their lack of professionalism. An examination of the structure of school districts reveals that one characteristic of educator's structure (teachers and principals) is their lack of autonomy within their own sphere of self-sufficiency. For example, teachers receive extensive training to be credentialed to teach. Their claimed areas of expertise (self-sufficiency) are content areas and child development. However, it is these areas that are directly affected every time a current program is to be implemented. To make matters worse, teachers are not normally the developers of the programs, but rather entrepreneurs and others less directly involved with children and instruction.

Mr Berry created a decentralized organization with self-sufficient units, teachers in their classrooms, principals responsible for their budgets and school programs in their schools, and a lean central office available for delivering critical services. Mr Berry, however, managed and led all of them in a systematic, uniform manner. He did this through the control of information. Contrary to the expectations of decentralization in the late 1960s and early 1970s, decentralization did not mean relinquishing power to external or internal groups. Neither did it mean permitting special-interest groups to articulate the functions and mission of the organization.

In the case of Mr Berry, decentralization meant that the function for which schools have existed in this country, that is, the provision of educational services for *all* children, could best be performed. Decentralization meant that each person and group engaged in the function of schooling needed to be nurtured, protected, and allowed to do its best. The superintendent controlled the total organization to insure this took place. The community and the school board, confident in Mr Berry's integrity allowed him to perform his job. Moreover, to insure that the intentions were not misconstrued, the superintendent worked on a year-by-year contract, alerting all to the absence of personal benefit considerations. This points to another critical characteristic of decentralization. Each self-sufficient unit is self-sufficient to the degree that it functions for the benefit of the whole organization, the school district, and not for personal or group gain. Overall control of the organization is necessary to insure that this is maintained.

March and Oleson (1976) write that 'organizations regulate connections among problems, choice opportunities, solutions, and energy by administrative practice. Hierarchy, specialization, the distribution of information, etc. are all devices for regulating the connections among the four streams' (pp. 31–32). 'Regulation of the four streams through intention, learning, and segmentation are not mutually exclusive. Segmentation emphasizes structural rights of participants, problems, solutions; learning and intention address the question whether choice opportunities are attractive for participants, problems and solutions' (p. 32). They conclude with 'organizational structure is a result of deliberate planning, individual and collective learning, and imitation' (p. 32).

The decentralization of the Riverside Unified School District meant that the segmentation of the organization was affected. This segmentation was regulated through the direct control of information. The structure was protected through instruction and intention and the school personnel viewed the process of decentralization as a period of challenge and excitement. This case is probably as excellent an example as can be presented that 'organizational structure is a result of deliberate planning, individual and collective learning and imitation' (March and Oleson, 1976, p. 32).

The above considerations are consistent with Allison's (1971) organizational process model in which goals, expectations and choice are the three major categories and the related variables are: quasi-resolution of conflict; uncertainty avoidance, problematic search and organizational learning (pp. 76–77). What is instructive in this particular case is that when Allison's (1971) organizational process model was applied to the various decentralization cases, the issues of segmentation, intention and learning were not primary. Decentralization was defined as the process imposed by the model rather than as process products as Levy (1966) presents, that is, self-sufficient units' autonomy within their own sphere of self-sufficiency.

The connection between decentralization and flight metaphors is the search for expanding the school boundaries as modifying the segmentation (see Hanson, 1984). This case demonstrates that a critical characteristic of decentralization is the support for organizational members' initiative in improving the organization.

Another implication is as Levy (1966) describes: 'The more highly interdependent the various elements in a relatively modernized society become, the less any administrative ingenuity in the form of decentralization of power can substitute for detailed adequate knowledge about the areas in which decision must be made' (p. 489). The decentralization in the Riverside Unified School District was both vertical and horizontal (see Mintzberg, 1974, p. 210). Here the decision power was concentrated largely among the teachers and principals as professionals engaged in a common endeavor. The coordination between schools throughout the district was conducted through the control of information. The protection of the professionals within this structure was likewise maintained through the control of information.

Conclusion

The distinction made between *theory of organization and theory of management* is not immediately obvious ... *Theories of organization* are seen as being concerned with all the components of an organization (e.g., a school) while *theories of management* are concerned largely with one domain of organization centering on authority, decision making etc. The *theory of change* is really a sub-section of management theory. Of interest in the present study is that its affinity has been more with theories of curriculum and their renewal than with management theory. Hoyle and McMahon (1986) claim that 'in the 1960s, it was optimistically assumed that organization theory would substantially enhance our understanding of schools' (see Hoyle, 1965). There is now a wide-spread view that this promise has not been fulfilled (see for example, Davies, 1982). The present study raises the issue that maybe we don't understand schools because in fact the schools have not been studied as *total* organizations. The research efforts have concentrated at the classroom and school level rather than at the district level. Aside from the decentralization studies, school district organizational studies are rare. As explained earlier, there has been a difference in the perceptions of teachers, principals and superintendents. There has also been a difference in the focus of interest of study. Researchers have been reluctant to view school districts as one complex organization engaged in delivering quality educational services throughout.

The definition of school improvement comes from Glatter (1986) who writes that it is 'a systematic sustained, effort aimed at change in learning conditions and other related internal conditions in one or more schools, with the ultimate aim of accomplishing educational goals more effectively' (p. 87).

Decentralization in Riverside was intentional, inclusive, and incremental. School reform took place in the Riverside Unified School District as an effort to improve the children's academic achievement and the staff's work performance. In the accomplishment of these goals, the organization was transformed from a bureaucratic organization to a decentralized one. The school reform effort originated and was carried out by the chief executive of the organization, the superintendent. The intentional acts of the superintendent resulted in decentralizing a medium-sized urban school district through seven stages. Although each of the stages are, in retrospect, theoretically distinguishable, they were not apparent to the participants during the process. The decentralization process included restructuring the central office, granting autonomy to principals and creating clusters for all school people to participate equally. Each of these acts held the potential for conflict.

The superintendent was successful in decentralizing the organization through the control of information. The superintendent instituted a research and evaluation unit, appointed an intellectual and a staff person to cover the various sorts of information. The research and evaluation unit served to process all student related data. The intellectual handled all information requiring theoretical, intellectual or academic expertise. The staff person processed all local, social and current information. Each of these nodes of information reported directly to the superintendent. The superintendent controlled the acquisition, processing and release of information from these three points.

What have we learned from this case study? Leadership style affects change processes. Intentional leadership is lodged in a person with a vision and sense of mission for the organization. The motivational basis is the common good for the organization rather than personal or special interest. The process by which the organization improves itself is as important as the improvement itself. The intentional leader instructs, trusts and experiments with the organizational members. Intentionalism includes a comprehensive view of the total organization as well as an appreciation for the personal and individual contributions of each member.

Successful organizational change is dependent on the sophisticated control of information rather than people. In schools, three critical points of information are under the control of the superintendent. The institutionalization of the research and evaluation unit insures that information regarding the mission of the school, instruction and children, is collected, processed and transmitted in a uniform, consistent and trustworthy manner. The presence of an intellectual insures that the organization is in constant contact with the academic community, has access to the latest theoretical and research information and has the capacity to have that information interpreted for district application. The post of the staff member in charge of current, local and social information assures the superintendent that all levels and sorts of information are being processed. The institutionalization of these three informational units provides the school executive with the capacity to obtain, process and transmit relevant information. Thus, the control of information serves as the technological core of the organizational leader. All decisions and actions are conditioned by the filtering process based on the core value, the common good, and the basic assumptions that children will learn, that those engaged in delivering educational services are responsible for what children learn, and that the quality of the participation of the organizational members matters. The effect of external forces is ultimately neutralized by the integrity of the value system and the sophistication of the technological core. This study demonstrates that successful, dramatic, and long-term organizational change takes place through the control of information.

The organizational change described in this study is one of transforming a bureaucratic organization into a decentralized one. Decentralization took place through seven stages which maximized intentionality and information control. The result is organizational units which contribute qualitatively to the organizational good. This allows organizational members to view decentralization as beneficial, challenging and personally gratifying. The overall organizational culture is likewise affected.

Chapter 7

Policy Implications

What have we learned from studying the leadership of Mr Berry during the period of decentralization of the Riverside Unified School District which may contribute to the improvement of schooling? First, the issue of school reform is to be lodged upon the office of the superintendency. Schools are organizations managed by persons, and as this study has shown, the superintendent can *intend*, that is, make choices as to how the school district will be managed. Thus, if schools are to improve, executive officers must assume the responsibility for doing those things which will bring about improvement. Practically, because there are only 16,000 superintendents in the US public schools, focusing on this office is more manageable than focusing on school sites.

Second, school reform must encompass the entire school organization which consists of a school district. Targeting separate schools for reform is not sufficient to bring about the school reform that is necessary to have social impact. Two aspects are related to this issue. First, principals and teachers at the school site level are subordinate to the superintendent's office. Expecting the lower ranks to perform independently has shown that successful schools tend to be rare.

Second, equality and excellence in schools will remain sporadic as long as school sites are focused upon. In a recent news release (Willis, 1987) the California Superintendent of Schools, Honig, relates that one reason a group of the 30 lowest performing high schools consisting primarily of minorities improved so much was because 'as part of (their) effort to hold the schools accountable for their students' performance, (Honig) personally contacted the superintendents of those school districts last year'. In this case, the realization that superintendents wield influence is presented. Thus, in order to maximize the delivery of excellent and equitable educational services, the superintendent must be held accountable at the district level. In this manner, individual schools will not be abandoned or favored unjustly.

Third, much concern has been expressed in the current reform reports regarding the inability to affect teachers' work performance. Training, reward systems, merit pay systems and other programs have been proposed. Our study shows that school personnel rise to the expectations of high work performance when the leader grants primacy to the quality of the participation as well as to the outcomes. In other words, the means are as important as the ends. Decentralization for the Riverside Unified School District was a challenging and exciting, as well as rewarding time for teachers and principals. In the case of schools, teachers, instruction and children are to be focused upon as contributing members rather than as vulnerable, weak, subordinate or victims. The Riverside Unified School District's overall school performance improved because teachers, instruction, and children were granted primacy. The school organization's accomplishments were subsequently highlighted rather than the organization's leader.

Fourth, the process of change in organization is a long one which requires a leader's continuous attention, from its inception to its completion. The case which has been reported demonstrates that leaders cannot be expected to drop into an organization, arouse its members to action and leave before the uncertainty of the change process has subsided and the learning of new roles and responsibilities has taken place. Two aspects are to be considered. First, the length of time it takes to complete a change process and second, using the impetus of organizational change starts for the leader's upward mobility. In both cases, the consequences are critical for the organization. Organizational changes are lengthy and incremental. Successful organizational change starts don't necessarily indicate successful school leadership. As has been illustrated in this case history, a successful leader is one who remains with the organization until its change is completed, rather than one who leaves in search of other interests.

References

ALLISON, G. (1971) *Essence of Decision: Explaining the Cuban Missile Crisis.* Boston: Little, Brown.

ALSCHULER, A.S. and IVEY, A. (1972) The human side of competency-based education, *Education Technology*, 12 (11), pp. 53–55.

ARGYRIS, C., (1973) Today's problems with tomorrow's organizations, in *Tomorrow's Organizations: Challenges and Strategies*, JUN, J. and STORM, W. (eds), Glenview, Illinois: Scott Foresman.

ARNEZ, N.L. (1981) *The Besieged School Superintendent: A Case Study of School superintendent-School Board Relations, in Washington, D.C., 1973–1975*, Washington, D.C.: The University of America Press.

ARNSTEIN, S. (1971) Eight rungs on the ladder of citizen participation, in *Citizen Participation: Effecting Community Change*, CAHN E. and PASSETT, B. (eds). Praeger Special Studies in United States Economics and Social Development, New York: Praeger.

ASTUTO, T.A. and CLARK, D.L. (1986) 'Achieving effective schools',: in *World Yearbook in Education, 1986: The Management of Schools*, edited by HOYLE, E. and McMAHON, A., New York: Kogan Page, London/ Nichols Pub. Co., pp. 57–72.

BACHARACH, S.B. and CONLEY, S.C. (1986) A managerial agenda for educational reform, *The Education Digest*, 52 (4) December, pp. 14–17.

BALDRIDGE, J.V. and DEAL, T.E. (1975) *Managing Change in Educational Organizations*, Berkeley, CA: McCutchan.

BALDRIDGE, J.V. and DEAL, T.E. (1983) *The Dynamics of Organizational Change in Education*, Berkeley, CA: McCutchan.

BARD, B. (1972) Is decentralization working? *Phi Delta Kappan*, 54 (4) pp. 238–43.

BARTLETT, H. (1957) More light on the core, *The Sociology of Education*, 21 (March), p. 106.

BERRY, E.R. (1971) *Report to the Board of Education on the Organization and Direction of the District*, Riverside, CA: Riverside Unified School District.

BLUMBERG, A. with BLUMBERG, P. (1985) *The School Superintendent: Living with Conflict*, New York: Teachers College, Columbia University.

BORMAN, K.M. and SPRING, J.H. (1984) *Schools in Central Cities: Structure and Process*, New York: Longman.

References

BOUTWELL, W.D. (1968) Happenings in education, *The PTA Magazine*, 63 (3) November, pp. 25–26.

BOWERS, C.A. (1984) *The Promise of Theory and the Politics of Cultural Change*, New York: Teachers College Press.

BRIGHT, R.L. (1967) The place of technology in educational change, *Audiovisual Instructor*, 12 (April) pp. 340–343.

BRITTENHAM, L.R.(1980) *An Ethnographic Case Study of the Administrative Organization, Processes and Behavior in an Innovative Senior High School*, Madison, Wisconsin: Wisconsin in Research and Development Center, June, ED 196 148.

BURNS, J.M. (1978) *Leadership*, New York: Harper and Row.

California Commission on the Teaching Profession (1985) *Who Will Teach Our Children? A Strategy for Improving California's Schools: The Report of the California Commission on the Teaching Profession*, Sacramento, CA: The California Commission on the Teaching Profession.

Carnegie Forum on Education and the Economy Task Force on Teaching as a Profession (1986) *A Nation Prepared: Teachers for the 21st Century: The Report Task Force on Teaching as a Profession*. Washington, D.C.: The Forum.

CAWELTI, G. (1974) Urban school decentralization and curriculum development strategies, pp. 18–27 in I.E. STAPLES (ed.) *Impact of Decentralization on Curriculum*, Washington, D.C.: Association for Supervision and Curriculum Development.

CHAPMAN, J. and BOYD, W.L. (1986) Decentralization, devolution, and the school principal: Australia lessons on statewide educational reform, *Educational Administration Quarterly*, 22 (4) Spring, pp. 28–58.

COLEMAN, J.S. (1966) *Equality of Educational Opportunity*, Washington, D.C.: US Office of Education.

COVEL, J.I. (1977) *Analysis of School Administrators' Careers in Riverside County from 1870–71, 1974–75*, Unpublished doctoral dissertation, University of California, Riverside, Riverside, CA.

CRUZ, J. (1985) Instructional leadership: A superintendent's contribution to school effectiveness, *Thrust*, 15 (3) November/December, pp. 14–16.

CUBAN, L. (1985) Conflict and leadership in the superintendency, *Phi Delta Kappan*, 67 (1) September, pp. 28–30.

DAVIES, B. (1982) Organizational theory and schools, in *The Social Sciences in Education Studies*, HARTNETT, A. (ed.), London: Heinemann.

DEAL, T.E. and KENNEDY, A.A. (1982) *Corporate Culture: The Rites and Rituals of Corporate Life*, Menlo Park, CA: Addison-Wesley Publishing Company.

DENZIN, N.K. (1977) *Sociological Methods: A Sourcebook*, New York: McGraw-Hill.

Economic Industrial Site Information (1965) Riverside, CA: Riverside Chamber of Commerce.

ETZIONI, A. (1961) *Complex Organizations: A Sociological Reader*, New York: Holt, Reinhardt & Winston, Inc.

FANTINI, M.D. (1972) Needed: Radical reform in schools to make accountability work, *Nation's Schools*, 89 (May) pp. 56–58.

FANTINI, M.D. and GITTELL, M. (1973) *Decentralization: Achieving Reform*, New York: Praeger.

FLETCHER, L. (1986) Education administration sharpened by the eyes of Janus: An historial approach to educational decision-making, *Journal of Educational Administration and History*, 18 (2) July, pp. 66–77.

FOLLETT, M.P. (1941) *Dynamic Administration: The Collected Papers of Mary Parker Follett*, METCALF, H.C. and URWICK, L., (eds), New York: Harper and Row.

FOSTER, N.J. (1983) *Toward a Theory of Teaching*, Unpublished Doctoral Dissertation, University of California, Riverside, California.

FULLAN, M. (1982) *The Meaning of Educational Change*, New York: Teachers College Press.

GEE, E.G. and SPERRY, D.J. (1978) *Education Law and the Public Schools: A Compendium*, Boston: Allyn and Bacon, pp. A–19 and A–20.

GERARD, H. and MILLER, N. (1975) *School Desegregation*, Chicago, Illinois: Plenum Press.

GIACQUINTA, J.B., (1973) The process of organizational change in schools, in *Research in Education*, KERLINGER, F.N. (ed.) New York: Peacock Publishers.

GITTELL, M. (1967a) *Participants and Participation: A Study of School Policy in New York City*, New York: Center for Urban Education.

GITTELL, M. (1967b) Problems of school decentralization in New York City, *Urban Review*, 2 (4) pp. 27–28.

GITTELL, M. and HOLLANDER, E.T. (1968) *Six Urban School Districts: A Comparative Study of Institutional Response*, New York: Praeger.

GLASMAN, N.S. (1986) *Evaluation-Based Leadership: School Administration in Contemporary Perspective*, New York: State University of New York Press.

GLATTER, R. (1986) The management of school improvement, in *World Yearbook of Education in 1986: The Management of Schools*, HOYLE, E. and McMAHON, A. (eds), New York: Kogan Page, London/Nichols Publishing Company, pp. 87–99.

GOODLAD, J.I. (1985) Can our schools get better? *Education Digest*, 51 (3) November, pp. 164–168.

Greater Riverside Chambers of Commerce, 4th edition (1983) Riverside, California: Sullivan Publications.

GROSS, N. and TRASK, E. (1976) *The Sex Factor and the Mangement of Schools*, New York: John Wiley and Sons.

GUTHRIE, J.W. and REED, R.J. (1986) *Educational Administration and Policy: Effective Leadership for American Education.* Englewood Cliffs, N.J.: Prentice-Hall, Inc.

HANSON, E.M. (1979) *Educational Administration and Organizational Behavior*, Boston: Allyn and Bacon, Inc.

HANSON, M. (1984) Exploration of mixed metaphors in educational administration research, *Issues in Education*, 2 (3) Winter, pp. 167–185.

HARNEY, R.F. (1979) *Oral Testimony and Ethnic Studies*, Ontario, Canada: The Multicultural History Society of Ontario.

HARRIS, B.H. (1972) Schools without subjects, *Educational Leadership*, 29 (February) pp. 420–3.

HEIDELBERG, R.L. (1971) Cafeteria Concept: Curriculum malnutrition, *Phi Delta Kappan*, 53 (November) pp. 174–175.

HENDRICK, I.G. (1968) *Development of a School Integration Plan in Riverside, California: A History and a Perspective*, Riverside, CA: Riverside Unified School District and University of California.

HERMANOWITZ, H.J. (1959) Problem solving as a teaching method, *Educational Leadership*, 18, pp. 299–306.

HERZBERG, F. (1966) *Work and the Nature of Man*, Cleveland: World Publishing Company.

HOLMES GROUP (1986) *Tomorrow's Teachers: A Report of the Holmes Group*. East Lansing, MI: The Holmes Group, Inc.

HOMANS, G.C. (1950) *The Human Group*, New York: Harcourt, Brace, Jovanovich and Company.

HOWE, H. II (1968) *Respect, Engagement, Responsibility*, Washington, D.C.: US Department of Health, Education and Welfare, Office of Education ED 020 989.

HOYLE, E. (1965) Organizational analysis in the field of education, *Educational Research*, 7 (2) pp. 97–114.

HOYLE, E. and MCMAHON, A. (eds), (1986) *World Yearbook of Education in 1986: The Management of Schools*, New York: Kagan Page, London/ Nichols Publishing Company.

HUGHES, L.W. and BRATTON, M.J. (1978) Perceived decision-making authority in a large decentralized urban school system, *Catalyst for Change*, 7 (3) Spring, pp. 4–9, 20.

IVEY, A.E. (1969) The intentional individual: A process-outcome view of behavioral psychology, *The Counseling Psychologist*, 1 (4) pp. 56–60.

KANTER, R.M. (1983) *The Change Masters*, New York: Simon and Schuster.

KNEZEVICH, S. (1975) *Administration of Public Education*, New York: Harper and Row.

KOWITZ, G.T. (1963) Examining educational innovations, *The American School Board Journal*, 146 (6) December, pp. 5–6.

LANDSMAN, T. (1962) The role of self concept in learning situations, *The High School Journal*, 45 (April) pp. 289–295.

LANOUE, G.R. and SMITH, B.L.R. (1973) *The Politics of School Decentralization*, Lexington, Mass.: Lexington Books.

LAWLER, E.E. (1986) *High Involvement Management: Participative Strategies for Improving Organizational Performance*, San Francisco: Jossey-Bass Publishers.

LAWSON, T.O. (1973) *Decentralization 1973: A Second-year Progress Report*, Los Angeles: Office of the Superintendent of Education and Management Assessment, Los Angeles Unified School District.

LAWSON, T.O., FORBES, W., COOGAN, J., CRANE, I., FITT, W., GRANT, R., LOPEZ, L., MCCULLOUGH, T. and SIMON, C., *Decentralization 1973: A Second-year Progress Report*, Los Angeles, CA: Los Angeles Unified School District, Office of the Superintendent, Education and Management Assessment (October) ED 091 822.

LEVIN, H.M., (1971) The case for community control of schools, in *New Models for American Education*, GUTHRIE, J.W. and WYNNE, E. (eds), Englewood Cliffs, N.J.: Prentice-Hall.

LEVY, M.J. JR. (1966) *Modernization and the Structure of Societies: A Setting for International Affairs*, Volumes 1 and 2. Princeton, N.J.: Princeton University Press.

LIPPITT, R. (1984) 'The preparation and nurturance of managers for today and tomorrow,' In *The Future of Management Education*, KABABADSE, A. and MUBBI, S. (eds), England: Gower Pub. Co., pp. 360–369.

LUTZ, F.W. and IANNACCONE, L. (1978) *Public Participation in Local School Districts*, Lexington, Mass.: Lexington Books.

MARCH, J.G. and OLESON, J.P. (1976) *Ambiguity and Choice in Organizations*, Bergen, Norway: Universitatforlaget.

MASLOW, A. (1954) *Motivation and Personality*, New York: Harper and Row.

MASTERS, N.A., SALISBURY, R.H. and ELIOT, T.H. (1964) *State Politics and the Public Schools*, New York: Alfred A. Knopf.

MAY, R. (1969) *Love and Will*, New York: W.W. Norton.

McEACHERN, G.M. (1968) Afro-American history: Schools rush to get in step, *Nation's Schools*, 82 (September) pp. 58–62.

McGIVNEY, J.H. and Haught, J. (1972) The politics of education: A view from the perspective of the central office staff, *Educational Administration Quarterly*, 8 (3) Autumn, pp. 18–38.

MEAD, G.H. (1934) *Mind, Self and Society*, Chicago: University of Chicago Press.

MENGES, R. (1977) *The Intentional Teacher*, Monterey, CA: Brooks/Cole Publishing Company.

MERCER, J.R. (1968) *Issues and dilemmas in school desegregation: A case study*, Paper delivered to the 17th Western Regional Conference on Testing Problems, San Francisco.

METCALF, F.D. and DOWNEY, M.T. (1977) *Teaching Local History: Trends, Tips, and Resources*, Boulder, Colorado: Social Science Consortium.

MILLER, L.S. (1986) The school reform debate, *The Journal of Economic Education*, 17 (3) Summer, pp. 204–209.

MINTZBERG, H. (1979) *The Structuring of Organizations*, Englewood Cliffs, N.J.: Prentice-Hall, Inc.

MOORE, M.T. (1975) *Local school program planning organizational implications*, Paper presented at The American Educational Research Association Annual Meeting, Washington, D.C. ED 105 593.

MORRIS, V.C., CROWSON, R.L., PORTER-GEHRIE, C. and HURWITZ, E. JR., (1984) *Principals in Action: The Reality of Managing Schools*, Columbus, Ohio: Charles E. Merrill Publishing Company.

National Commission on Excellence in Education (1983) *A Nation at Risk: The Imperative for Education Reform*, Washington, D.C.: US Government Printing Office.

National Governors' Association (1966) *Time for Results: The Governors' Report on Education*, Washington, D.C.: National Governors' Association.

ORNSTEIN, A. (1974) *Metropolitan Schools: Administrative Decentralization vs. Community*, METUCHEN, N.J.: The Scarecrow Press.

ORTIZ, F.I. and HENDRICK, W. (1987) A comparison of leadership styles and organizational culture: Implications on educational equity, *Journal of Educational Equity and Leadership*, 7 (2) Summer.

OUCHI, W. (1981) *Theory Z: How American Business can Meet the Japanese Challenge*, Reading, Mass.: Addison-Wesley Publishing Company.

PARKS, E.C. (1959) Adequate education for each American youngster, *National Association of Secondary-Principals*, 43 (October) pp. 98–101.

PATTERSON, T. (1971) *A Colony for California*, Riverside, California: Press Enterprise Company.

PATTERSON, J.L. and HANSEN, L.H. (1975) Decentralized decision-making: It's working in Madison, *Educational Leadership*, 3 (2) November, pp. 126–128.

PERROW, C. (1965) Hospitals, technology, structure and goals, in *The Handbook of Organizations*, MARCH, J.G. (ed.) Chicago: Rand McNally.

PETERS, T.J. and WATERMAN, R.H., JR. (1982) *In Search of Excellence: Lessons From America's Best-run Companies*, New York: Harper and Row.

PETERSON, P.E. (1976) *School Politics Chicago Style*, Chicago, Illinois: University of Chicago Press.

PILO, M.R. (1974) *Sequential and organizational models of school decentralization: New York City and Detroit*, Paper presented at the American Educational Research Association Annual Meeting, Chicago, ED 094 030.

Press-Enterprise Microfilm Film, 1960–1978 (1960–1978) University of California, Riverside, California.

Public Schools Directory for Riverside County (1960–1979) California, Riverside, California: Riverside County Schools Office.

RAMOS, A.G. (1981) *The New Science of Organizations*, Toronto, Canada: University of Toronto Press.

RAMSEY, C.P. (1963) Current issues in programmed instruction, *Education*, 83 (7) March, pp. 412–415.

RAVITCH, D. and GRANT, W. (1975) *School Decentralization in New York City 1975 and Detroit's Experience with School Decentralization*, Washington, D.C.: Center for Governmental Studies, ED 113 428.

Report Card, Volume 1, Number 8, (1977) Riverside, California: Riverside Unified School District.

RICE, J.M. (1893) *The Public School System of the United States*, New York: Century.

Riverside City Hall (1983) *Historical Information*, Riverside, California: Mayor Brown's Office, September.

Riverside County Department of Development (1983) *Community Economic Profile for Riverside, Riverside County, California*, Riverside, CA: Riverside County Department of Development.

Riverside County Registrar of Voters (1983) *Information Registration*, Riverside, CA: Registrar's Office.

Riverside Press-Enterprise (1982) Bruce Miller, Ex-RUSD Head, dies, *Riverside Press-Enterprise* (September 15) p. B-1.

Riverside Unified School District (1965) *Bulletin Board*, 12 (2) November.

Riverside Unified School District (1979) *Documents on Cluster Evaluation*, Riverside California: Riverside Unified School District (May 16) pp. 8–10.

Riverside Unified School District (1960–1979) *Minutes of the Board of Education of the Riverside Unified School District*, Riverside, California: Riverside Unified School District.

ROGERS, D. (1968) *110 Livingston Street: Politics and Bureaucracy in the New York City Schools*, New York: Random House.

ROGERS, D. and CHUNG, N.H. (1983) *110 Livingston Street Revisited: Decentralization in Action*, New York: New York University Press.

ROWAN, B. (1981) The effects of institutionalized rules on administrators, in *Organizational Behavior in Schools and School Districts*, BACHRACH, S.B. (ed.) New York: Praeger.

SCHEIN, E.H. (1985) *Organizational Culture and Leadership*, San Francisco: Jossey-Bass Publishers.

SCHLECHTY, P.C. and JOSLIN, A.W. (1986) Images of schools, in *Rethinking School Improvement: Research, Craft and Concept*, LEICHMAN, A. (ed.), New York: Teachers College, Columbia University, pp. 147–161.

SCHUTZ, A. (1967) *The Phenomenology of the Social World*, Evanston, Indiana: Northwestern University Press.

SCHWEBECK, M. (1969) Everyone can be educated if the system is changed, *Nation's Schools*, 83 (May) p. 10.

SIZEMORE, B.A. (1981) *The Ruptured Diamond: The Politics of the Decentralization of the District of Columbia Public Schools*, Washington, D.C.: University Press of America.

SPROULL, L.S., WEINER, S., WOLF, D.B. (1978) *Organizing an Anarchy*, Chicago, Illinois: University of Chicago Press.

State of California, (1977) 'Title V, Chapter 7, Plans to alleviate racial and ethnic segregation of minority students', *State of California, California Administrative Code*, Sacramento, CA: Office of Administrative Procedure. (Filed, 9-16-77).

STEINBERG, L.S. (1974) *Participation and representation in an age of decentralization and alternatives*, Paper presented at the American Educational Research Association Annual Meeting, Chicago, ED 089 417.

STEINBERG, L.S. (1977) *Social Science Theory and Research on Participation and Voluntary Associations: A Bibliographic Essay*, New York, N.Y.: Optimum Computer systems, Incorporated, ED 144 869.

STEINBRUNER, J.D. (1974) *The Cybernetic Theory of Decision: New Dimensions of Political Analysis*, Princeton, N.J.: Princeton University Press.

TYACK, D.B. (1974) *The One Best System: A History of American Urban Education*, Cambridge, Mass.: Harvard University Press.

US Bureau of the Census (1967) *County and City Data Book*, Washington, D.C.: US Department of Commerce Publication, p. 477.

US Bureau of the Census (1972) *County and City Data Book*, WASHINGTON, D.C.: US Department of Commerce Publication, pp. 474–653.

VERGERONT, E. (1971; revised 1972) *Read and Listen*, Riverside, CA: Riverside Unified School District.

WEBER, MAX (1957) *The Theory of Social and Economic Organization*, Glencoe, Illinois: The Free Press.

WEDEL, C.C. (1966) God and the new math, *International Journal of Religious Education*, 43 (December) pp. 6–7.

WEICK, K.E. (1979) *The Social Psychology of Organizing*, Reading, Mass.: Addison-Wesley.

WILLIS, D. (1987) High school test scores hit 10-year high, *Riverside Press-Enterprise*, Tuesday, March 17, pp. A1 & A4.

WINTER, G. (1966) *Elements for a Social Ethic*, New York, N.Y.: Macmillan Co.

WISSLER, D.F. (1984) *The Decentralization of Decision-making in Riverside Unified School District: An Historical Analysis,* Unpublished doctoral dissertation, University of California, Riverside.

WISSLER, D.F. and ORTIZ, F.I. (1986) The decentralization process of school systems: A review of the literature, *Urban Education*, 21 (3) October, pp. 280–294.

ZIEGLER, H., KEHOE E. and REISMAN, J. (1985) *City Managers and School Superintendents: Response to Community Conflict*, New York: Praeger.

ZEIGLER, L.H., PIERCE, L., SONNENFELD, D., WEBB, R., GROVE, J. (1973) *The Responsiveness of Public Schools to Their Clientele: Milestone 1: Report of Progress*, Eugene, Oregon: Center for the Advanced Study of Educational Administration, ED 127 699.

Author Index

Allison, G. 3, 138, 156
Alschuler, A.S. and Ivey, A. 15
Arnez, N.L. 121, 139
Arnstein, S. 71

Baldridge, J.V. and
  Deal, T.E. 154
Bard, B. 138
Bartlett, H. 28
Berry, E.R. 45, 68, 103, 107, 109
Blumberg, A. and Blumberg, P.
  120, 121, 122, 141
Borman, K.M. and Spring, J.H.
  120, 138, 140
Boutwell, W.D. 28, 29, 30
Bowers, C.A. 16
Bright, R.L. 29
Brittenham, L.B. 51, 57, 73, 78
Burns, J.M. 27, 51, 78

Cawelti, G. 38, 81
Chapman, J. and Boyd, W.L.
  38, 82
Coleman, J. 24
Cuban L. 141
Covel, J.I. 143, 144, 145

Davis, B. 157
Deal, T.E. and Kennedy, A.A. 1
Denzin, N.K. 11

Etzioni, A. 3

Fantini, M.D. and Gittel, M. 13,

27, 30, 111, 113, 117
Fletcher, L. 31
Follett, M.P. 98
Foster, N.J. 15
Fullan, M. 152

Gee, E.G. and Sperry, D.J. 125
Gerard, H. and Miller, N. 5, 19,
  23, 25, 26, 115, 147
Giaquinta, J.B. 110, 111, 115
Gittell, M. 1, 81, 113, 116, 129, 138
Gittell, M. and Hollander,
  E.T. 112
Glasman, N.S. 123, 127, 135
Goodlad, J. 2
Governor's Commission, 1
*Greater Riverside Chamber of
  Commerce* 21, 142
Guthrie, J.W. and Reed, R.J. 125

Hanson, M. 156
Hendrick, I.G. 146, 147, 148
Holmes, E.W. 143
Hoyle, E. 157
Hoyle E. and McMahon, A.
  142, 157
Hughes, L.W. and Bratton, M.J.
  138

Ivey, A.E. 15, 32

Kanter, R.M. 4, 125, 135, 154
Knezevich, S. 120
Kowitz, G.T. 29

Subject Index